NATURAL KINDS, LAWS OF NATURE AND SCIENTIFIC METHODOLOGY

AUSTRALASIAN STUDIES
IN HISTORY AND PHILOSOPHY OF SCIENCE

VOLUME 12

The titles published in this series are listed at the end of this volume.

NATURAL KINDS, LAWS OF NATURE AND SCIENTIFIC METHODOLOGY

Edited by

PETER J. RIGGS
Department of Philosophy, and
Department of Physics and Mathematical Physics,
University of Adelaide,
Australia

KLUWER ACADEMIC PUBLISHERS
DORDRECHT / BOSTON / LONDON

A C.I.P. Catalogue record for this book is available from the Library of Congress

ISBN 0-7923-4225-9

Published by Kluwer Academic Publishers,
P.O. Box 17, 3300 AA Dordrecht, The Netherlands.

Kluwer Academic Publishers incorporates
the publishing programmes of
D. Reidel, Martinus Nijhoff, Dr W. Junk and MTP Press.

Sold and distributed in the U.S.A. and Canada
by Kluwer Academic Publishers,
101 Philip Drive, Norwell, MA 02061, U.S.A.

In all other countries, sold and distributed
by Kluwer Academic Publishers Group,
P.O. Box 322, 3300 AH Dordrecht, The Netherlands.

Printed on acid-free paper

Printed in the Netherlands

TABLE OF CONTENTS

FOREWORD

Australia and New Zealand boast an active community of scholars working in the field of history, philosophy and social studies of science. 'Australasian Studies in History and Philosophy of Science' aims to provide a distinctive publication outlet for their work. Each volume comprises a group of essays on a connected theme, edited by an Australian or a New Zealander with special expertise in that particular area. In each volume, a majority of the contributors is from Australia or New Zealand. Contributions from elsewhere are by no means ruled out, however, and are indeed actively encouraged wherever appropriate to the balance of the volume in question. Earlier volumes in the series have been welcomed for significantly advancing the discussion of the topics they have dealt with. The present volume will I believe be greeted equally enthusiastically by readers in many parts of the world.

R. W. Home
General Editor
Australasian Studies in History
and Philosophy of Science

ACKNOWLEDGEMENTS

The stimulus to assemble this collection of essays grew from a number of interesting seminars conducted during the academic years 1991–92, under the auspices of the Victorian Centre for the History and Philosophy of Science (which is centred at the Bundoora campus of La Trobe University in the outer suburbs of Melbourne). Some of the seminar presentations to the Victorian Centre for HPS and other articles awaiting publication have been made available as part of the Pre-print Series of the Victorian Centre. All of the papers in this volume were invited with the intention of assembling them into a single collection .

I wish to thank the General Editor of the Australasian Studies series, Professor Rod Home, for his assistance in getting this volume approved for publication. Other thanks must go to the anonymous referees who kindly provided their time and expertise for what usually is a thankless job; Ms. Annie Kuipers and Ms. Evelien Bakker of the Humanities and Social Sciences Division of Kluwer Academic Publishers; Professor Graham Nerlich of the Department of Philosphy, University of Adelaide, who made my family feel very welcome in Adelaide and was a strong source of encouragement; Dr. Alistar Blake, Head of the Department of Physics and Mathematical Physics, University of Adelaide, for the use of departmental facilities in the production of this volume and Mr. Mark Ferraretto for assistance with data processing.

PETER J. RIGGS

PETER J. RIGGS

INTRODUCTION: NATURAL KINDS AND OTHER ASSORTED CONCEPTS

This volume of the Australasian Studies in History and Philosophy of Science is devoted to the topics of natural kinds, scientific methodology, laws of nature and their many and varied interconnections. One can readily point to such connections, for instance: our scientific theories tell us about the sort of physical entities we can expect the world to consist of (be they natural kinds or other entities); through the application of appropriate scientific methods (e.g. induction) we postulate/discover laws of nature; where two or more scientific theories compete for acceptance, a choice is made with reference to appropriate methodological rules; and so on. We find, therefore, a very broad expanse of related issues. Paul Churchland succinctly summarised the resulting juncture of philosophical subjects surrounding the notion of natural kinds when he wrote:

> The problem of natural kinds forms the busy crossroads where a number of larger problems meet: the problem of universals, the problem of induction and projectibility, the problem of natural laws and de re modalities, the problem of meaning and reference, the problem of inter-theoretic reduction, the question of the aim of science, and the problem of scientific realism in general.[1]

Each contribution to this volume highlights one or more aspects of the interconnecting themes.

Although the subjects of natural kinds and laws of nature have been written about and discussed extensively,[2] recent work has brought out important associations with other areas of philosophy. One example is a paper by William Harper,[3] who draws together induction and natural kind reasoning. He argues that inductive generalizations made on the basis of small amounts of data are legitimate when made with respect to a conception that indicates what features are essential to the kind in question. Another example is a article by John Macnamara,[4] in which it is argued that the criteria for the application of natural kinds to individuals result from a process of induction. The essays in this volume extend further the scope of these associations and also offer new insight into some of the traditional problems in philosophy of science. The essays themselves, are divided into four main groups: (1) Natural Kinds, Laws and Necessity; (2) Induction and Belief; (3) History and Methodology of Science; and (4) Specific Cases of Theories and Laws.

NATURAL KINDS, LAWS AND NECESSITY

The idea that there are certain kinds to be found in (or defined by) nature can be traced back at least to the time of Aristotle. The traditional natural kinds are familiar as the classifications of chemical elements and biological species. The idea of natural kinds has evolved beyond this traditional picture though and now may be said to include such entities as elementary particles. What exactly are natural kinds? How are they defined? Do they really exist or are they just a convenient conception for categorising the natural world, i.e. are there natural kinds or only artificial kinds? If natural kinds do exist, what sort of properties do they have? Such questions need to be answered if we are to take seriously any ontological commitment to natural kinds.

Various philosophers at different stages of history have argued for (and some against) the existence of (variously defined) natural kinds. John Venn, best known for the Venn Diagram, gave an apt description of natural kinds in 1866:

> ... large classes of objects, throughout all the individual members of which a general resemblance extends ... Such regularity as we trace in nature is owing, much more than is often suspected, to the arrangement of things in natural kinds, each of them containing a large number of individuals.[5]

The logician Willard Quine thinks that kinds are sets and that the notions of similarity and kind appear substantially the same.[6] However, to define natural kinds merely as sets seems somewhat inappropriate. A more generally accepted definition of natural kinds is "functionally relevant groupings in nature".[7] Yet none of these definitions capture the full flavour of the more contemporary views on the character of natural kinds where the notion of the essence of a kind has come to the fore. The following definition by Wilkerson is a better expression of this feature of natural kinds:

> Something is a member of a natural kind if and only if it has a real essence, an intrinsic property or set of properties that make it the kind of thing it is, irrespective of any system of classification ...[8]

Essentialism is a position which asserts that objects have essential properties (or essences) without which these objects could not be what they are.[9] In the theory of Scientific Essentialism, natural kinds play a central role. The principal tenets of Scientific Essentialism are as follows:[10]

(i) The world consists ultimately of things belonging to fundamental natural kinds.

(ii) The most fundamental natural kinds are distinguishable by their intrinsic properties and structures.

(iii) The intrinsic properties and structures which distinguish natural kinds from each other are their essential properties.

(iv) The essential properties of the most fundamental natural kinds are dispositional.

Reflection on natural kinds and essentialism conjures up a whole host of additional philosophical 'baggage', in particular concepts of *necessity*. The necessity referred to here is not analytic or logical necessity, but is of a different variety. The term 'necessity' usually is taken to mean an absence of exception. This is a useful definition for it finds a place not just in logic but in conceptions of natural law, e.g. the law of gravity is, in some sense, necessary as there are no exceptions to it. There are various notions of (non-logical) necessity found in the philosophical literature. The ones most commonly alluded to are: *natural necessity* and *physical/causal necessity*. Natural necessity (broadly interpreted) is the thesis that there are necessary connections in nature. The terms 'physical necessity' and 'causal necessity' are used (somewhat) synonymously, although not all philosophers follow this. Whilst the denial of something that is logically necessary produces a contradiction, the denial of something that has physical/causal necessity is supposed to constitute a violation of one or more laws of nature. There also is the modality of *metaphysical necessity*. This is defined using the apparatus of possible worlds — a proposition is metaphysically necessary if it is not an *a priori* proposition and is true in all possible worlds. Of course, not everyone agrees that these modalities really exist.

If natural kinds exist, then laws of nature would be ultimately tied up with them, but how and to what extent? What about the natural laws themselves, are they necessary, contingent or (as philosophers such as D.M. Armstrong assert) *contingently necessary*? The answers to these and other questions about the physical laws depends crucially on what theory of laws is subscribed to. The (once popular) Regularity theory of laws holds that laws of nature merely are regularities found in the behaviour of things, e.g. if all As are Bs then it just happens that anything that is an A is also a B. However the Regularity theory is no longer considered satisfactory for a number of reasons,[11] not the least of which is that accidental generalizations do not support counterfactual conditionals whereas laws do. A strong current trend is to account for laws of nature in terms of relations between universals, as L.J. Cohen has noted:

> In recent years a number of philosophers have come to advocate the thesis that natural laws are (second-order) relations of a certain kind between (first-order) universals or properties.[12]

The relationship between natural kinds and universals is examined, *inter alia*, in this section of the volume.

John Collier defines natural kinds as classes of objects that are related by laws of nature. In a short essay entitled 'On the Necessity of Natural Kinds', Collier affirms that natural kinds are pivotal to most systematic reasoning and especially to our reasoning about nature. He holds the view that generalities supervene on particulars and argues that both laws and natural kinds supervene on particulars. Natural laws depend on natural kinds as laws express necessary relations between kinds of individuals. Therefore a law is instantiated if the relevant kind exists, and in turn, the kind only exists if the particular exists. Since one aim of science is to find natural laws, Collier argues that these may be found by discovering that individuals of a kind exist.

Brian Ellis has a solid commitment to the existence of natural kinds. This derives from a desire to explain the complex structure and relations in the natural world. In 'Natural Kinds and Natural Kind Reasoning', Ellis puts forward a conception of natural kinds in order to describe the essential features of the world as revealed by scientific investigation. He argues that we need an ontology of both natural properties and natural kinds. Natural kinds are defined as objectively existing objects that can be grouped naturally according to their structures and properties such that these groups will be distinct ontologically. Natural kinds, however, are not merely property classes (which are classes of things that share a particular characteristic). In Ellis' theory, natural kinds have structured hierarchies and since property classes do not, natural kinds cannot be property classes nor classical universals. Neither are natural kinds intersections of property classes.[13] Natural properties are defined as the causal powers, propensities and capacities of the most fundamental kinds of things in the world, but natural properties do *not* define natural kinds. The causal laws of science are defined to be statements which describe the essences of the natural kinds of causal processes and Ellis argues that their modal status is that of metaphysical necessity. In regard to the aims of science, we are told that one basic aim is the discovery of what natural kinds of things exist and what their essential natures are.

In 'The Jerrybuilt House of Humeanism', Caroline Lierse presents a *bold* and *controversial* argument in which she specifically criticises what she describes as the 'Humean' positions of David Lewis and D.M. Armstrong. Lierse's argument centres around Hume's Distinctness Thesis which states that there cannot be any necessary connections between distinct existences. Lierse's major claim is that "... the unhappy fact is that they [the 'Humeans'] do abandon the Distinctness Thesis and embrace a causal criteria in one form or another ... this prohibited metaphysical baggage is never explicitly declared ... [and] infiltrates the whole Humean philosophical edifice."[14] Consequently, she argues that the various metaphysical 'edifices' which the 'Humeans' have erected (such as Lewis's Principle of Plenitude and Armstrong's Combinatorialism) are contradictory. A short reply by D.M. Armstrong may be found at the end of the volume.

Peter Forrest in his 'Physical Necessity and the Passage of Time' addresses a problem that remains outstanding in the philosophy of time, i.e. accounting for the apparent passage (or flow) of time. Forrest presents an otherwise unexplained hypothesis (called the 'thesis of the plurality of actual worlds') in order to explain both the passage of time and physical necessity, justifying this by claiming that progress is made if several mysteries can be reduced to only one. This theory is not one of possible worlds (*á la* David Lewis' modal realism) but of a single universe (world) in which there are many ways that it actually *is*!

What is necessary is taken to mean that which is beyond being affected. Therefore the past (to us) is necessary but was not so when it was future. The world at a given time *t* and all that is earlier than that *t* is actual. But so too is anything that is physically necessary after *t*. Necessity is characterised as that which is true in all actual worlds. In this theory, actuality is a time-dependent concept. What's more, actuality is cumulative. Thus there is not so much a passage of time as there is a growth into time, i.e. there is more that is actual at later times.

INDUCTION AND BELIEF

Inductive inference has a distinguished place in the history of scientific methodology, dating back to the time of the ancient Greeks. In Western intellectual circles, Sir Francis Bacon was considered as the 'Father of the Scientific Method' due to his published account of scientific research, the heart of which was a form of sophisticated induction. David Hume formulated what is commonly known as the (traditional) problem of induction, although Hume never labelled it so. This can be stated simply as the problem of how to move from a finite set of observation statements about a phenomenon to a general (i.e. universal) statement. Hume thought that induction was indispensable in our daily lives, but concluded that it could not be rationally justified. When an induction is made, it always is based on what is referred to loosely as the 'Principle of Induction'. This is a principle employed to justify inductive inference. Statements of this principle vary, but a generally accepted form is that "the future will be similar to the past". Nelson Goodman highlighted the difficulties that arise when applying the Principle of Induction in his 'New Puzzle of Induction' (or simply Goodman's Puzzle/Riddle). This puzzle, as is well known, concerns making inductive inferences about (artificial) properties named 'grue' and 'bleen'. Both essays in this section have something to say about Goodman's Puzzle.

The name of John Clendinnen has come to be associated closely with efforts aimed at showing the legitimacy of inductive inference. In the first volume of the Australasian Studies series, Clendinnen argued that inductively based predictions are rational because they may succeed, whilst predictions based on other criteria are not rational because they require an arbitrary decision.[15] In 'Theorizing and Empirical Belief', Clendinnen continues his justification of inductive inference by building on his earlier arguments. Although Clendinnen would agree that rules which determine when one has sufficient data to justify making an induction in any possible situation cannot be formulated, he does argue that a satisfactory explication of induction is essential to obtaining standards by which we can establish the rationality of our beliefs about the external world. Indeed, Clendinnen is of the opinion that induction cannot be isolated from our acceptance of perceptual beliefs for in order to predict we must identify patterns in empirical data and this requires that there be some trust in our perceptions.

Moreover, these patterns should cover the widest possible range of experience. The projection of objective patterns first needs some criteria by which basic properties may be identified. Clendinnen suggests ways of establishing these criteria, offers a solution to Goodman's Puzzle, and argues for an essential role for theory in the systematizing of our empirical beliefs.

A principle that sometimes arises in the context of induction is the Principle of Indifference. This states that no event should be considered as being any more probable than another event if there does not appear to be a reason why one event should occur rather than any other.[16] Critics of this principle are quick to point out that lack of knowledge does not justify acceptance of an indifferent attitude, or for that matter, any other attitude! In 'Indifference and Induction', Len O'Neill offers a rationale for the Principle of Indifference and attempts to show how two objections against the Principle may be overcome. In defending the Principle of Indifference, O'Neill places himself in the camp of those who would argue that "[w]e reason ... from an island of knowledge in a sea of ignorance."[17] O'Neill's rationale (a minimax loss argument) is one that justifies 'betting odds' and not degrees of confidence. The relevant objections are those from *epistemological magic* and the *non-linearity problem*. The former is a claim that it is not reasonable to produce degrees of belief out of a state of ignorance. The latter relates to the production of information in units that have non-linearly related magnitudes which, in turn, leads to different outcomes. The formal similarity between the problem of non-linearity and Goodman's Puzzle also is examined.

HISTORY AND METHODOLOGY OF SCIENCE

The history of science has provided much material to substantiate claims for or against the existence of proper methodological principles. Almost every philosopher of science uses historical evidence to support his or her own conclusion about scientific practice and scientific methodology. A prime example is the work of Paul Feyerabend who claims that there is an absence of methodological standards in science (i.e. "anything goes").[18] One of Feyerabend's favourite cases is that of Galileo who maintained an adherence to the Copernican system contrary to the physics of the day and in the face of apparently refuting empirical facts. (See also Thomason's essay below.)

William Whewell was both a philosopher and historian of science of the mid-1800s, whose theory on the nature of science seems to have not received the consideration which was its due. Whewell developed procedures for the conduct of science from his detailed study of the history of science.[19] Whewell paid tribute to the insights on scientific method of Francis Bacon, especially for his use of the process called the Colligation of Facts.[20] This Colligation also was an integral part of Whewell's method which he believed would lead to true theories. Briefly, Whewell's position was that science was a process in which ideas and 'facts' are linked together. This is achieved by making an inductive inference which requires a conceptual step by the scientist who "superinduces" a conception onto a set of 'facts'.[21] Further generalizations are gained through a consilience of inductions. Whewell described this procedure in the following terms:

> The Consilience of inductions takes place when an Induction, obtained from one class of facts, coincides with an Induction, obtained from another different class. This consilience is a test of the truth of the Theory in which it occurs.[22]

Whewell's method is sometimes referred to as *Hypothetico-Induction*, in stark contrast to the more familiar Hypothetico-Deduction. Whewell believed that his method would yield true scientific theories, although it is clear from his writings that he did not believe experience alone could show a proposition to be necessary. However, he seemed to think that insight provided by intuition could establish some theories as *necessarily true*.[23] It is unfortunate that Whewell's ideas about scientific methodology and of (what he called) natural classes has not been given much attention even in contemporary debates on methodology.[24]

In 'Towards Metamethodology: For the History and Philosophy of Science', John Fox asks why Whewell's method has not been followed and discusses aspects of Lakatos' methodology in the context of an overall examination of *metamethodological* issues.[25] Fox utilises the thesis denoted as *Scientism* due to F.A. von Hayek in his book *The Counter-Revolution of Science*.[26] Scientism (not to be confused with scientology) is a form of 'methodological and epistemological monism' which claims that science and mathematics are to be taken as paradigms for good methodological practice in other fields. Fox advocates an account of restricted Scientism plus Whewell's methodology.

Neil Thomason takes another look at Galileo's explanation of the moons of Jupiter in 'Elk Theories – A Galilean Strategy for Validating a New Scientific Discovery'. One of the criticisms which Galileo had to deal with was that the telescope was unreliable when viewing the celestial realm (a criticism which had its basis in the Aristotelean world-view of that time). The case of Galileo's observations and his inferences on the matter has been a much disputed episode in the history of science. As mentioned above, some commentators have employed it in order to draw strong conclusions about the lack of a consistent methodology in science and the irrationality of the whole enterprise. In a previous essay, Thomason used ARCHED hypotheses (ARCHED is an abbreviation for *Advanced Reinforcing Chancey Hypotheses Explaining Data*[27]). These are a set of hypotheses which individually are poorly supported but collectively are mutually supportive, like the two sides of an arch. Thomason argued, contrary to Feyerabend, that Galileo was rational. In the current essay, Thomason presents us with a different argument specifically targeted at the question of telescopic dependability, based upon what he has named *Elk Theories*. An Elk theory is one that has minimal content, so much so that just about everyone will agree with it. Thomason argues that an Elk theory of the telescope was engaged by Galileo for the purposes of validating the telescope as a reliable instrument for observing celestial objects. Again, contrary to Feyerabend, Thomason concludes that Galileo was successful in this exercise. Keith Hutchison comments on the power of Elk Facts at the end of this volume.

SPECIFIC CASES OF THEORIES AND LAWS

In this section the essays are devoted to issues arising out of our two best physical theories: the General Theory of Relativity and Quantum Theory. The advent of General Relativity in 1915 had the effect of enormously enriching our perceptions of reality and provided philosophers and physicists with new ontological conceptions to grapple with, such as the possible existence of physical non-Euclidean space. However, the inception of Quantum Mechanics in the 1920s with its Positivistic overtones has resulted in much dissatisfaction with the theory's orthodox interpretation, which seemed to deny the existence of an objective, observer-independent reality. Orthodox Quantum Theory also is plagued with a number of paradoxes which stubbornly refuse to go away. This situation has led to (at least) a dozen different interpretations of the theory being made, with each interpretation having its own way of dealing with the various paradoxes.

Let's quickly review some of the relevant details of the orthodox interpretation of quantum mechanics (also called the Copenhagen Interpretation). A quantum system is described by a state vector in a mathematical (Hilbert) space. It is assumed that the state vector contains all information about the state that can be obtained experimentally. If left to itself, a quantum system is said to be in a *superposition of eigenstates* and its time evolution is governed deterministically. However, a measurement of some physical quantity of the system, e.g. momentum, projects (or collapses) the state vector into *one* of the eigenstates of its superposition and this eigenstate is what is observed. This is a process that is inherently non-deterministic and for which the theory only can predict the *probability* that a particular value for the physical quantity will be found on measurement. This dual evolution of a quantum system leads to various paradoxes, such as the famous story of Schrödinger's Cat. In this tale, the cat is in a sealed box and its life or death depends on a (random) quantum event happening, such as a radioactive decay. If the event occurs, the cat dies. If it does not occur, the cat lives. We do not know the result until we make a measurement (e.g. look into the box). Orthodox quantum theory tells us that upon observation the cat's state vector is projected into either the eigenstate where the cat is alive or the eigenstate where it is dead, but that prior to the observation the cat is in a *superposi-*

tion of live and dead states! This is, of course, absurd. Such are the quantum paradoxes and the orthodox interpretation sheds precious little light upon them.

Adrian Heathcote offers some thoughts on the laws and interpretation of quantum theory in his 'Of Crows and Quarks: Reflections on the Laws of Quantum Mechanics'. What are the laws of quantum mechanics? The answer is not as straight forward as one might at first think. Heathcote identifies the whole mathematical structure of quantum mechanics as constituting its laws which is quite different to the identification of laws in classical physics. In regard to the nature of scientific laws themselves, Heathcote argues that the Humean conception of laws as natural regularities is incorrect, as is inductive scepticism. In addition, he rejects the opinion that laws are relations between unconditional properties (universals) as this does not mesh well with the quantum mechanical description of the micro-world. Instead Heathcote argues that natural laws are relations between *states* of all kinds (and this includes classical states as a subclass). Heathcote also presents some pertinent comments about the problem of measurement in quantum theory.

In 'Laws and States in Quantum Mechanics', John Forge seeks to add credence to the instance view of explanation. This is a view where to explain something is to show it to be an instance of some pattern in nature. In doing so, Forge embraces Bas van Fraassen's Modal Interpretation of Quantum Mechanics with the explicit aim of utilising the notion of physical laws as relations between states which are defined to be collections of properties. In the Modal Interpretation there are two sorts of 'states' present: the dynamical state and the value state. The dynamical state is what would be called the state of the system in the orthodox interpretation, i.e. described by state vectors. Its evolution is deterministic. The value state is more like the states found in classical physics — it tells us what observables possess values and what these values are.[28] Under this scheme, there is no need for two dynamics and no 'collapse' upon measurement. This avoids the problems generated by having two different types of evolution. Yet, despite using the interpretation developed by an anti-realist (van Fraassen), Forge commits himself to a form of realism for quantum mechanics by insisting that states have an existence independent of observers. Adrian Heathcote comments on Forge's essay at the end of this volume.

Ian Hinckfuss takes an unashamed nominalist stance when it comes to theories about the nature of space, i.e. a position requiring that theories which entail the existence of abstractions (e.g. numbers, points) should be considered only as instrumental. In 'Instrumental Theories: Possibilities and Space and Time', Hinckfuss argues that instrumentalism can be a viable option to scientific realism. He justifies this stance by claiming that only substances (and not abstractions) give rise to causation. Since the gaining of knowledge is a causal process, he infers that knowledge can only be about substances. Yet, we do speak of abstractions as if they are real. Hinckfuss defines the term: *presuppositions* to be elements of discourse (suppositions) that generally are present and are used as if were true. In other words, presuppositions are abstractions that are spoken of as if they exist. This essay is concerned with the removal of such presuppositions, a process that Hinckfuss labels — *presuppositional eliminativism.* Central to the argument is the notion of a conservative extension — this is the presupposition of an extension to a theory that performs the function of simplifying the theory by introducing an artificial ontology. Hinckfuss illustrates his ideas by providing a case study on the instrumentalism of geometry.

In 'Spacetime or Quantum Particles: The Ontology of Quantum Gravity?', I examine (in a particular case) what results when theories/laws from one domain are applied to a different domain. The case in point is Quantum Gravity, which should account for situations where both gravity and quantum mechanical effects are significant. At the time of writing, no complete theory of Quantum Gravity exists. In relativity theory there is a phenomenon known as time dilation, where time intervals between events are not equal in different frames of reference. A fairly well known example is the asymmetric ageing of twins. I argue that the gravitational version of the asymmetric ageing of identical biological specimens shows that (geometrically interpreted) curved spacetime is not dispensable. This result is used to conclude that the particle-based interpretations of quantum gravity are not acceptable.

University of Adelaide, South Australia

NOTES

[1] Churchland (1985), p.1.
[2] Examples include: the Simon Fraser Conference on Natural Kinds, February 1982; French *et al.* (eds) (1986).
[3] Harper (1989), pp.144-46.
[4] Macnamara (1991).
[5] Venn (1964), p.55.
[6] Quine (1969), p.8.
[7] Hacking (1990), p.130.
[8] Wilkerson (1986), p.64.
[9] Flew (1979), p.112.
[10] Ellis (1992), p.1.
[11] Armstrong (1983), Chap. 2.
[12] Cohen (1987), p.16.
[13] See: Ellis and Lierse (1994).
[14] Lierse (this volume).
[15] Clendinnen (1982).
[16] Flew (1979), p.171.
[17] O'Neill (this volume).
[18] Feyerabend (1975).
[19] Whewell (1857).
[20] Whewell (1860), p.135.
[21] Losee (1980), p.124.

[22] Whewell (1858), Aphorism XIV, Chap.5.
[23] Butts (1965), p.165.
[24] The prominent exception has been the work of Robert Butts.
[25] For relevant background, see: Riggs (1992), pp.12-13, 181.
[26] von Hayek (1952), pp.15-16.
[27] See Thomason (1994).
[28] Forge (1993), p.15.

REFERENCES

Armstrong, D.M. (1983) *What is a Law of Nature?* Cambridge: Cambridge University Press.

Clendinnen, F. J. (1982) 'Rational Expectation and Simplicity' in McLaughlin, R. (ed.). *What? Where? When? Why? Essays on Induction, Space and Time, Explanation. Australasian Studies in History and Philosophy of Science Volume 1*. Dordrecht: Reidel.

Butts, R.E. (1965) 'Necessary Truth in Whewell's Theory of Science', *American Philosophical Quarterly* **2**, 161-81.

Churchland, P.M. (1985) 'Conceptual Progress and Word/World Relations: In Search of the Essence of Natural Kinds', *Canadian Journal of Philosophy* **15**, 1-17.

Cohen, L.J. (1987) 'Laws, Coincidences and Relations between Universals' in Pettit, P., Sylvan, R. and Norman, J. (eds.), *Metaphysics and Morality: Essays in Honour of J. J. C. Smart*. Oxford: Blackwell.

Ellis, B.D. (1992) 'Scientific Essentialism'. *Victorian Centre for the History and Philosophy of Science Preprint Series 1/92*. Bundoora: La Trobe University.

Ellis, B. D. and Lierse, C. (1994) 'The Fundamental Importance of Natural Kinds'. *Victorian Centre for the History and Philosophy of Science Preprint Series 3/94*. Bundoora: La Trobe University.

Feyerabend, P. (1975) *Against Method*. Verso Press, London.

Flew, A. (ed.). (1979) *A Dictionary of Philosophy*. London: Pan.

Forge, J. (1993) 'The Modal Interpretation of Quantum Theory'. *Victorian Centre for the History and Philosophy of Science Preprint Series 3/93*. Bundoora: La Trobe University.

French, A.P., Uehling Jr., T.E. and Wettstein, H.K. (eds) (1986) *Studies in Essentialism. Midwest Studies in Philosophy XI*. Minneapolis: University of Minnesota Press.

Hacking, I. (1990) 'Natural Kinds' in Barrett, R.B. and Gibson, R.F. (eds), *Perspectives on Quine*. Oxford: Basil Blackwell.

Harper, W. (1989) 'Consilience and Natural Kind Reasoning' in Brown, J.R. and Mittelstrass, J. (eds). *An Intimate Relation: Studies in the History and Philosophy of Science*. Dordrecht: Kluwer.

Losee, J. (1980) *A Historical Introduction to the Philosophy of Science*. Second edn. Oxford: Oxford University Press.

Macnamara, J. (1991) 'Understanding Induction', *British Journal for the Philosophy of Science* **42**, 21-48.

Quine, W.V.O. 'Natural Kinds' in Rescher, N. (ed.), *Essays in Honor of Carl G. Hempel. A Tribute on the Occasion of his Sixty-Fifth Birthday*. Dordrecht: Reidel, 1969. Reprinted in Quine's *Ontological Relativity and Other Essays*. New York: Columbia University Press, 1969; and in Boyd, R., Gasper, P. and Trout, J.D. (eds), *The Philosophy of Science*. Cambridge, MA.: MIT Press, 1991.

Riggs, P.J. (1992) *Whys and Ways of Science: Introducing Philosophical and Sociological Theories of Science*. Melbourne: Melbourne University Press.

Thomason, N. (1994) 'The Power of ARCHED Hypotheses: Feyerabend's Galileo as a Closet Rationalist', *British Journal for the Philosophy of Science* **45**, 255-64.

Venn, J. (1964) *The Logic of Chance*. Fourth edn. New York: Chelsea.

von Hayek, F.A. (1952) *The Counter-Revolution of Science*. Glencoe, Ill.: Free Press.

Whewell, W. (1857) *The History of the Inductive Sciences*.

Whewell, W. (1858) *Novum Organum Renovatum.*
Whewell, W. (1860) *On the Philosophy of Discovery.*
Wilkerson, T.E. (1986) 'Natural Kinds and Identity, A Horticultural Inquiry', *Philosophical Studies* **49**, 63-69.

JOHN COLLIER

ON THE NECESSITY OF NATURAL KINDS *

Natural kinds are central to most scientific reasoning about the world. For that matter, they are central to most kinds of systematic reasoning that are not purely analytic. In this essay I will investigate why natural kinds are used in science and the extent to which science requires them. These issues revolve around the role of nomological necessity in science and the role of natural kinds in natural laws. I will give an account of nomological necessity and the necessary (essential) properties of natural kinds that distinguishes the necessity involved from analyticity, logical necessity, and metaphysical necessity. The distinction is based on a metaphysical difference between propositions that must be true and propositions that *cannot* be false.

Systematic reasoning about the world requires the classification of particular objects and properties found in the world so that generalisations can be made about them. The simplest form of systematic statement about the world is a correlation of the extensions of two classes of such particulars. More complex statements that relate classes of particulars can be formulated purely extensionally, but modal concepts are required to go beyond mere correlations. Systematic reasoning about the world need not go beyond mere correlations. We might decide to restrict systematisation just to correlations (perhaps because of epistemic scruples about unobserved entities), or else it might be that a particular world we want to reason about does not support anything but correlations (for example, a world in which all correlations are accidental). If, however, we have reason to include general statements supporting counterfactuals (laws) in our systematic reasoning, we must also permit modal relations.

There are three general reasons for thinking that we need to support counterfactual statements in scientific reasoning. First, scientific laws and principles support counterfactuals. That is, if a person accepts a scientific law, then, *ceteris paribus*, that acceptance sanctions their asserting particular counterfactual statements of the same form as the law.[1] (The main exception would be cases in which the person accepts the law, but is considering situations in which the law might or does not hold.) This reason is a very weak one, since it is purely descriptive of scientific reasoning in practice. It does not tell us that science needs counterfactuals, unless we have some additional reason for thinking that something that has laws that do not support counterfactuals would not be science. It would not gain us anything to make this a matter of the definition of science.

A second reason for thinking that scientific reasoning must support counterfactual statements is that it uses inductive techniques to establish universal generalisations. In science, under appropriate conditions we project observed regu-

P.J. Riggs (ed.), Natural Kinds, Laws of Nature and Scientific Methodology, 1–10.

larities to all future cases. That is, in projecting to some presumed law that all As are Bs, we infer that if we were to find an A, then it would also be a B. The subjunctive form of the inference indicates its support for counterfactuals. It might appear that this gives us our additional reason for thinking that science needs counterfactuals since science, at the very least, requires inductive generalisations to universal statements. There is a way around this conclusion, however. We can interpret scientific laws purely extensionally, in which case particular instances would be material conditionals. We can then explain the subjunctive reasoning above as reducible to material conditionals. The subjunctive form, however convenient, is not required.

This move, however, must be suspect, since it seems to be central to science to distinguish between accidental generalisations and scientific laws. The latter, in addition to being simply true, involve some form of necessary connection, reintroducing modalities. Since scientific laws are contingent, the necessity involved must in some way depend on the particularity of our world. Thus it cannot be metaphysical necessity or analyticity, since both of these are, in different ways, independent of the particularity of our world. The most likely candidate is the sort of necessity associated with causation. This brings us to the third reason for thinking that scientific reasoning requires counterfactuals: the centrality of causal reasoning to scientific reasoning.

Causal reasoning is common in science. Despite attempts to expunge causality from science, it has been notably resilient. Scientific explanations are irreducibly causal in their nature, even though there has been a long history of attempts to replace causal with inferential or statistical relations.[2] Causal reasoning invokes counterfactual statements by establishing correlations among hypothetical classes.[3] Necessity plays a central role in causal reasoning, since ascertaining causal connections by direct observation, abstraction and comparison alone is impossible, as David Hume firmly established.

I suppose that a science involving only correlations would be possible. It would be justified if there were nothing but accidental correlations in the natural world. However, if some correlations are non-accidental, but still contingent, then they fall clearly within the domain of science. Causal relations, in particular, fit into this category. Since our world is one in which causation occurs, then science, at least contingently, must support counterfactuals.

A causal relation between a and b, such that if a occurs, b must occur, requires that a is of such a kind that if it occurs, something of a kind of which b is a member must occur. That is, there must be two kinds, one of which a is a member and the other of which b is a member, such that if an event of the first kind occurs, then an event of the second kind must occur. Many of the characteristics of a and b will be incidental to their causal interaction. We can strip these away by abstraction until we are considering only those properties that are required to maintain the necessity of the causal interaction. We are left with the properties defining the least determinate classes to which a and b must belong to in order to maintain the necessity of their causal interaction. Anything which instantiates these two classes will also have a causal connection (due to the maintenance of necessity). Thus, the relation between these least determinate classes constitutes a universal necessary relation. I would be inclined to call this a law, except for one problem. Particular values of the general properties are (in general) required for a to cause b (imagine a is an initial configuration of particles, and b is a later configuration). Laws, on the other hand, apply to a range of values.

Tempting as it is to abstract away from the particular values of properties essential to the causal interaction of a and b to reach the realm of universals, we cannot do this and still guarantee causation will occur. Not just any values for the properties of a and b essential to causation will do. In order to preserve causation, we need to preserve the mathematical relations between the relevant properties of a and those of b. To abstract away from the particular values instantiated in a and b so that the universal necessary relation abstracted to is of the form of a law, L, the mathematical form of L must be incorporated into the properties defining the appropriate classes for a and b. This is required in order to preserve the necessity of causal relations that are instances of L. We can conclude that abstraction from causal instances to causal laws entails that the mathematical form of the law is implicit in the properties essential to the causal interaction in the instances. Otherwise, abstraction would not be solely to less determinate properties, i.e., it would not be a generalisation. So, the abstraction of general causal laws from particular causal instances entails that properties involved in the particular instances contain all of the entailments of the properties involved in the laws. Furthermore, the necessity of laws supervenes on the necessary relations of particulars. It also seems reasonable to hold that the laws themselves supervene on the relations of particular things; nothing that makes a difference is gained by holding otherwise. Nonetheless, we can say that laws are necessary relations between kinds of things,[4] the kinds being defined as abstractions from the values of the properties of the least determinate class required to maintain the necessity of the causal relations of the things that instantiate the laws.[5] A law, then, will hold, and hold necessarily, if it is instantiated at least once.

To recap briefly: A modestly rich science must presume that there are discoverable necessary connections in the world. Causal relations are the most familiar of these, if not the sole example. Assuming that causal relations hold in virtue of certain properties of the related entities, these properties must imply the causal laws they support, i.e., if the properties required to support the necessity of causation are instantiated, then the law itself is instantiated. No additional conditions (e.g., that the properties are instantiated in a world in which the law holds) are required. This follows from the stipulation that the relevant properties are abstractions from the concrete causal properties that ensure the necessity of causation in particular instances. On this account, necessitation in particular instances is primitive. I will show below that the extension of this necessity to classes and laws by abstraction gives an adequate account of nomological necessity.

It is first necessary to show that these classes are natural kinds. Since the classes related by laws are causally based, and are derived ultimately from causal relations in the natural world, it seems reasonable to call them natural classes in the sense of Duhem.[6] According to Duhem, a natural classification is one that classifies things according to the way they are related in the world. Such a classification might be more or less abstract, depending on the degree of determination of the classifying properties. Highly abstract classes are of little use to science, since they are difficult to use for prediction, control and explanation. The most useful classification is into kinds, such that if something falls into a kind, the type of its relations to things of other kinds are fully determined (i.e., as determined as they can be). Duhem would probably deny the scientific relevance of classifications of this sort, because they would seem to him to involve metaphysics (essential relations). If we include causal relations under the scope of science, however, causal kinds (and other kinds involving contingent necessitation) would form the basis for such useful classifications. Since their basis is entirely natural, it is reasonable to consider them as candidates for natural kinds. These kinds correspond directly to the natural classes I have been describing. They are fundamental, in the sense that no less determinate class would do the job (causal relations would not be preserved), but also that no more determinate class would do the job (it would be unnecessarily specific). I propose, then, to identify the classes I have defined with natural kinds, and I conclude that natural kinds are (contingently) required by science.

What else can be said about these natural kinds? First, the set of natural kinds gives the most general natural classification sufficient to account for causal necessity in particular cases. This follows from the abstraction process by which they were defined. Birds and bees are not natural kinds, since: (1) their causal interactions with other things can be abstracted to more general classes; and (2) biologically, their origin depended on particular circumstances involving particular values of certain causal properties. The first point, and perhaps the second, applies as well to the chemical elements. Whatever the natural kinds are, they must be very abstract. Perhaps masses and intervals are natural kinds. Second, necessary relations (other than analytic and metaphysical relations) among natural kinds are given by the laws that relate them. This follows from the need to incorporate the laws into the definitions of the kinds. Third, the existence of a natural kind implies that these natural necessary relations hold, at least potentially. It is impossible for the relevant kinds to exist, and for these relations not to hold. Again, this follows from the requirement that the mathematical relations asserted by laws be incorporated into the definitions of the kinds. Fourth, because of the requirement of maximal indeterminateness, there is a unique classification into natural kinds. If there are two different routes along which to abstract from instances, both of which maintain the necessity of the causal interaction, then neither is least determinate. At the very least, a disjunction of the endpoints of the two would be less determinate, but would still maintain the required necessity.

There are several reasons against taking the classes defined above as natural kinds. First, it might appear that least determinacy is too strong a condition, since laws need not relate fundamental entities. The gas laws, for example, describe gases, but gas is not a natural kind on my account, since gases are (a) made up of particles and (b) under certain conditions cannot be distinguished from liquids and/or solids. The best response to this complaint is that not all laws are fundamental, and evidence for the non-fundamentality of a law is evidence that the basic instances have either not been identified correctly or have not been abstracted to greatest generality, or both. Even non-fundamental laws, however, would still follow from the necessary relations of natural kinds, since the non-fundamental laws are particular consequences of the fundamental laws. Second, it might be objected, the kinds obtained from the proposed definition are not the traditional natural kinds. The simplest answer to this objection is that science has been wrong before, and will likely be shown wrong again. Third, it might be objected that there is no guarantee that the defined classes will form any useful groupings at all. They might be too diffuse, and many could be singular. The simplest answer in this case is to allow the possibility, and remark that it is fortunate for science that we are not in a world in which the possibility holds too widely. A fourth set of objections hold that the defined classes do not meet certain *desiderata* for natural kinds. I will deal with these next.

It would be desirable to show that if something belongs to a natural kind, it belongs to it essentially, that nothing can belong to inconsistent natural kinds, and further to show that nothing can belong to more than one natural kind. The first is impossible to show in general, since it is possible to define highly gerrymandered entities. Likewise, the second and third points have counter examples among the arcana of mereology. If we are going to satisfy the three desiderata above, we must restrict our scope to natural entities. Anything must have a unity relation that makes its parts, parts of the same thing. For most, if not all, natural things, this will be a causally based relation. Elsewhere, I have called this causally based relation cohesion.[7] In general, cohesion has a spatial and a temporal component. The cohesion of a thing is essential to it; destroy the cohesion, and you destroy the thing. It seems reasonable (I can think of no exceptions) that the same basic laws underlie the cohesion of a thing throughout its existence.[8] If so, then there will be at least one natural kind involved throughout its existence, and that natural kind will be essential to it. This alone does not preclude the possibility that the thing also belongs inessentially to some natural kinds, but if we are serious about saying that the thing as a whole, and not some part of it, interacts causally with something else, it seems likely that the properties involved in the kind or kinds it belongs to essentially will be involved in its causal interactions with other things. If so, any kind it belongs to, it would belong to essentially. Again, I can think of no exceptions. An electron, for example, can interact through its mass, spin, charge or size, but it cannot do without any of these and still be an electron.

If my argument that natural things belong to the natural kinds they belong to essentially is close to correct, then nothing can belong to inconsistent natural kinds, even at different times. All the natural kinds a thing belongs to, it must belong to at all times. On the other hand, objects can be of more than one natural kind, or so it seems. An electron, for example is both charged and massive, but some particles are just massive. So far as we now know, electromagnetism and gravitation are different properties, so it seems that electrons belong to both the kind gravitating system and the kind electromagnetic system. Properties, though, because of the abstracting process to least determinate property, cannot belong to more than one kind unless there are different routes of abstraction to a most general property that can sustain the necessity of causal interactions in which they are involved. I can think of no cases. Present evidence suggests, then, that natural kinds give a unique natural classification that is unitary for properties, but not for objects.

Despite the above, there are still some reasons to believe that scientific classifications are always accidental, even if loaded with the language of necessity. If so, necessity plays no real role in science, and natural kinds are at best arbitrary. One possibility is that even if there are natural classifications, they are inaccessible to science; any classification we can make is completely or largely accidental to any properties the things classified might actually have. Several arguments have been directed towards this being systematically true, but aside from some reflectively dubious empirical generalisations by some historians and sociologists of science, it seems to me that they all stem from the inseparability of theory and observation. I have argued that elsewhere that the point behind the theory ladenness of observation is a dilemma for any empiricist.[9] Either observed entities are rich enough to provide an empirical basis for theorising (are projectible), in which case they are pre-classified in potentially questionable ways, or else they are not pre-classified, in which case they are not rich enough to provide an empirical basis for theorising. If we cannot unequivocally determine what properties are involved in an instance of causation, then we can hardly abstract from them to a unique classification, as described in the earlier part of this essay. The same point, I believe, underlies Nelson Goodman's new riddle of induction, and arguments for Kuhn-Feyerabend incommensurability and Quinean indeterminacy of translation. I believe the problem can be met, but to do so would be outside of the scope of this essay. Instead, I will set such epistemological problems aside in order to examine the nature of the necessity involved my account of natural kinds.

The traditional distinction between universals and particulars suggests that necessity (if invoked) must be ontologically fundamental, since relations among particulars depend on the existence of the particulars, and are thus contingent (though they may be necessary given the existence of the particulars). David Lewis, however, has developed a systematic view of the world in which all generalities, including necessary generalities, are supervenient on particulars.[10] The cost is a commitment to ontologically fundamental possible worlds, and to a view of chance that is a little bit peculiar. My account avoids these problems (the latter I cannot show here), but at the expense of making the necessity of particular instances of causation primitive. Otherwise, I agree here that all generalities supervene on particulars. The differences won't matter in what follows.

Scientific laws must be contingent, but it is unclear how the same thing could be both necessary and contingent. We need an account of necessity that allows for the sort of contingency that we have come to expect in science. The way I have defined natural kinds, if something behaves as it does in virtue of its natural kind, it could not behave otherwise. This might seem unduly restrictive. Indeed, it might seem to imply that the necessity involved in natural kinds as I have defined them is really metaphysical necessity. This appearance is wrong, as I will show, because of my reliance on the supervenience of the general on the particular. It will be useful to first look at other alternatives for reconciling contingency with necessity.

One place to look is in the function of language. Analyticity is a form of necessity that is distinct from metaphysical necessity, the contrary of contingency. Perhaps laws can be analytic but still contingent. Scientific laws often do have some similarities to analytical statements. For example, Newton's Second Law not only states an empirical fact, but it also helps to define the notions of mass and force within Newtonian mechanics. The Second Law of Thermodynamics plays a similar role in defining entropy. Poincaré has argued that a number of laws have conventional aspects, and function like definitions.[11] The necessity of these laws would follow from their analyticity, but they would be contingent, it might be held, because they do not have to be accepted if their linguistic context is rejected. This is analogous to sentences like "I am here now." This sentence is true whenever uttered (barring unusual meanings given to the words involved), yet it does not state a necessary truth.[12]

That there is something wrong with this account of the necessity of laws should be evident from the distinctly unlaw-like character of "I am here now." Also, laws like Boyle's Law and Charles' Law do not seem to be in the least analytic. How can we explain their necessity, consistent with their contingency? The relevant form of necessity is not analyticity. An analytic statement must hold by meaning alone, but scientific laws are, presumably, empirical, synthetic statements. Any contingency involved in analytic statements derives from the circumstances of use of the terms or concepts involved. The contingency of scientific laws, however, is largely independent of how they are stated or conceived.

Another possible approach is to note that the necessity of laws holds only in worlds that are nomologically similar to ours. Presumably this would be the case for other necessary but contingent relations among natural kinds as well. There are two problems with this approach. First, without a clear criterion of nomological similarity, the approach merely reproduces the problem of defining the appropriate sort of necessity. It is unclear how to compare nomological similarity. E.g., do exceptions to laws count? If so, how much? Second, unless the properties corresponding to natural kinds exist only in the nomologically similar worlds, the same property could both be and not be a natural kind, depending on world. It is impossible to give this objection much more flesh without a satisfactory account of nomological similarity across worlds.

The worlds approach does suggest a solution to the problem of nomological necessity, however. Laws hold in all and only those worlds in which the appropriate natural kinds exist. There will be worlds in which a given law does not hold, because the appropriate natural kinds do not exist. Thus it is contingent that a given law holds. However it is impossible for the law to be false; in this sense it is necessarily true. On this account there is no problem with the same properties determining natural kinds in one world, but not in another. We can eliminate possible worlds talk in this account by revising the description of the position to say that if a natural law were not to hold, then the natural kinds on which it depends would not exist (the contrary is not necessarily true).

It might appear that this approach to nomological, and hence causal, necessity is unduly restrictive for individual causal events. It appears that they could not be other than they are. This suggests that any causal event we are involved in, we must be involved in. While it is true that in so far as the causally relevant factors are concerned, any causal event could only happen as it did, this places little restriction on what can happen. It is still possible that other events could have happened instead, either like or dislike the ones that actually happened. Furthermore, the aspects of an event that are not essential to the necessity of causation could have been different in any case. It is true that a tight web of causal events might allow only one outcome, but that sort of determinism is not peculiar to the present account of nomological necessity.

It is important to be clear how the supervenience approach to necessity differs from essentialist accounts that place the contingency of laws only in the kinds of things that happen to exist in this particular world. On the essentialist account, natural laws are metaphysically necessary, but may hold vacuously. The biggest problem with this approach, other than the ontological exuberance of vacuous laws and natural kinds, is that metaphysical necessity implies that the laws should be discoverable a priori. No particular experience is required. This seems to run deeply against the grain of scientific methodology.

On my account, however (and also Lewis's account), since natural kinds and laws supervene on particulars, there is no need to postulate metaphysical necessity as well to explain the necessary relations of natural kinds. Laws hold if and only if the relevant kinds exist, and the relevant kinds exist if and only if the relevant particulars exist. Furthermore, we must discover the particulars in order to discover the laws, since the laws are abstractions from the relations of particulars.[13]

We discover natural kinds first by discovering that individuals of the kind exist, and then by determining that individuals of that kind bear certain necessary relations to those of other kinds, in virtue of their being members of the kind they are. Thus natural kinds are doubly empirical. They are contingent in so far as their existence is concerned, as required for the contingency of science, but they bear necessary relations that include the empirical scientific laws. It is this contingency that underlies the contingency of scientific laws. Because of the nature of the necessity of these laws, they cannot be false, but they do not necessarily need to be followed.

University of Newcastle, New South Wales

NOTES

* I would like to thank the anonymous referee for useful comments on an earlier draft.

1 Although I use the term 'laws' here, everything I say can be generalized to theories that cannot be expressed as a simple set of conditional laws, capable of being applied individually. The translation uses a model-theoretic approach, such that if a particular system satisfies a particular theory, then it will have such-and-such properties.

2 E.g. Salmon (1984, 1989).

3 Giere (1984).

4 I have been deliberately ambiguous about whether the things I am talking about are objects or properties of some sort. Since objects are individuated solely by their properties, I am not convinced it makes much difference. See below for further considerations.

5 If necessary connections other than causation are (contingently) required by science, the notion of law can be amended to include them.

6 Duhem (1954), p.25.

7 Collier (1988).

8 The Aristotelian principle that something must remain constant through a change in anything goes in the right direction, but is not itself adequate to get this result.

9 Collier (1987).

10 Lewis (1994).

11 Poincaré (1905).

12 Extensive discussion of cases like this, together with explanation of the analyticity involved and the sense of metaphysical necessity are given by Kaplan (1979) and by Stalnaker (1976, 1978).

13 Generally, but not always by any means, the epistemological process will be more convoluted.

REFERENCES

Collier, J. (1987) 'Theory Ladenness' in P. Weingartner and G. Schurz (eds.), *Logic, Epistemology and Philosophy of Science: Proceedings of the 11th Annual Wittgenstein Congress* (Vienna, Holder-Pichler-Tempsky).

— (1988) 'Supervenience and Reduction in Biological Hierarchies' in M. Matthen and B. Linsky (eds.) *Canadian Journal of Philosophy (Supplementary Volume)* 14, *Philosophy and Biology:* 209-34.

Duhem, P. (1954) *The Aim and Structure of Physical Theory* (Princeton: Princeton University Press).

Giere, R. (1984) *Understanding Scientific Reasoning* 2nd edition (New York, Holt, Reinhart and Winston).

Kaplan, D. (1979) 'On the Logic of Demonstratives', *Journal of Philosophical Logic* 8, pp.81-98.

Lewis, D.K. (1994) 'Chance and Credence: Humean Supervenience Debugged', *Mind* 103: 473-90.

Poincaré, H. (1905) *Science and Hypothesis* (London, Walter Scott).

Salmon, W. (1984) *Scientific Explanation and the Causal Structure of the World* (New York, Princeton University Press).

— (1989) *Four Decades of Scientific Explanation* (Minneapolis: University of Minnesota Press).

Stalnaker, R. (1976) 'Propositions' in A. MacKay and D. Merrill (eds.) *Issues in the Philosophy of Language* (Yale: Yale University Press).

— (1978) 'Assertion' in P. Cole (ed.) *Syntax and Semantics* 9: *Pragmatics* (Academic, N.Y.).

BRIAN ELLIS

NATURAL KINDS AND NATURAL KIND REASONING*

It is at least plausible to suppose the world consists fundamenatally of natural kinds of substances, properties and processes. But whether this is so or not, it is certainly the case that a great deal of scientific research is devoted to their study. Most chemical research, for example, is concerned with the study of natural kinds of substances and processes, and with the attempt to discover their essential natures. Much of fundamental physics is about natural kinds of particles or fields, the kinds of properties they display, and the kinds of processes in which they may be involved. Even in biology, there are fundamental kinds of structures and processes which deserve the name 'natural kinds'. It is important, therefore, to develop a theory of natural kinds adequate to reflect these concerns, to investigate the logic of natural kinds, and to elaborate a theory of natural kind reasoning. This essay attempts to make a start on this project.

I shall proceed to describe the natural kind ontology that I think is needed for science, and the reasons why such an ontology is required. However, it will not be possible, in the space of this essay, to justify this ontology fully, since this would require an essay at least as long as the present one.[1] I shall then develop the concepts of intrinsicality and essentiality required for natural kind reasoning, and explore the logic of natural kinds. Finally, I shall make some observations about natural kind reasoning, and outline a theory about the nature of such reasoning.

1. WHY NATURAL KINDS ARE NEEDED IN ONTOLOGY

The old Aristotelian ontology of properties and relations (universals) and their bearers (particulars) is still very widely accepted. However, it is not obvious that everything can be accounted for in these simple terms, and philosophers have long struggled, without notable success, to develop satisfactory analyses of laws, dispositions, quantities, modalities, and the like, compatible with this basic ontology.

Moreover, this traditional metaphysic has become increasingly out of step with scientific knowledge. For modern science is at least largely about describing the essential natures of the natural kinds of objects, properties and processes which can occur. Therefore, we need a basic ontology which can accommodate such kinds of things. The classical ontology of universals and particulars would appear to be inadequate for this task. What seems to be needed is a more so-

P.J. Riggs (ed.), Natural Kinds, Laws of Nature and Scientific Methodology, 11–28.

phisticated metaphysic which has an ontology of *natural kinds* as fundamental physical things in the world.

Basically, the trouble with the traditional ontology is that it cannot account satisfactorily for all of the structure and necessary connections which we find about us. If the world consists just of properties and relations, and the individuals which instantiate them, then how can we explain the existence of hierarchies of natural kinds of objects, and the necessary connections between their properties and modes of behaviour? What kind of account can we give of quantities, which are evidently not simple properties, but which, nevertheless, may be as fundamental as anything else in nature? And how can we explain the existence of hierarchies of natural kinds of processes, which would appear to have no place at all in the traditional ontology, but which are, nevertheless, the basis of our natural laws?

To account for all this structure, we need a more sophisticated ontology — one which includes various categories of *natural kinds* as fundamental entities. Traditionally, natural kinds are just natural property classes, i.e. classes of things which share a certain complex natural property, such as that of being a horse, or of being an electron. However, there is good reason to think that natural kinds are not just natural property classes[2] and are therefore not ordinary universals.[3] They are, of course, like ordinary universals in some ways. But they are also importantly different from them. Moreover, natural kinds are not just the intersections of natural property classes.[4] Of course, if they were, then natural properties would be ontologically more fundamental than natural kinds. But this is not so either. Natural kinds are as fundamental in ontology as natural properties.

A second shortcoming of traditional ontology is that it does not readily accommodate natural kinds of events or processes. Yet, science textbooks are full of discussions of naturally occurring processes. Pick up any elementary physics text, and you will find descriptions of such natural kinds of processes as refraction, diffraction, reflection, electromagnetic radiation, and so on. But, sadly, metaphysics has all but ignored their existence. The main reason for this would appear to be that no theory of processes naturally falls out of the traditional metaphysic. Nevertheless, natural kinds of processes surely do exist, and their study is a very important part of the scientific enterprise. Think of the processes of biological reproduction, i.e. of meiosis and mitosis. An understanding of these processes is just about as fundamental in biology as anything can be. It is surely one of the basic aims of natural science, in biology, as well as in the physical sciences, to discover what kinds of fundamental natural processes can occur, and what their essential natures are.

A third shortcoming of traditional ontology is that it does not readily accommodate quantities. Aristotle had a separate category for them. It has no difficulty with specific quantitative properties, such as a length of two metres, or a mass of four grams. But a quantity is not just the class of its specific quantitative instances, since the same quantity would exist whatever its specific instances might be, and the spectrum of a continuous quantity does not have any gaps, even if there happens to be no instance anywhere of some specific quantitative property which would fall within this spectrum. Most plausibly, continuous quantities belong to a class of natural kinds which we may call 'spectral kinds'.[5] These are natural kinds whose instances are specific quantitative properties, and whose possible instances belong to a continuous spectrum of such properties. An ontology which allows such entities as spectral kinds has no difficulty in accommodating quantities.

For all of these reasons, natural kinds are needed as a basic category in ontology. They are not more fundamental than the objects, substances, properties or processes which instantiate them. But nor, as we have argued elsewhere,[6] are they just property classes, or the intersections of such classes. I am agnostic whether natural kinds can, like Platonic Forms, exist uninstantiated, or whether, like universals (on many accounts), they can only exist if they have instances. For the purposes of this essay, it does not matter which is the case.

2. NATURAL KINDS AND PROPERTY CLASSES

Often, things are said to be the same which are intrinsically different. They are classified together because they affect us similarly, (e.g. the use of the same colour terms to describe transmitted, reflective and radiant colours), or serve some common purpose, (e.g. in highlighting or illuminating). On the other hand, things are often distinguished from each other in ways which do not correspond to any real distinctions in nature (e.g. the distinctions between adjacent colours or species). The distinctions are made according to how we perceive things, or what we take to be distinguishing characteristics, although ontologically the distinctions we make may be arbitrary.

To explain any supposed similarities or differences which depend on the sorts of beings we are, e.g. on how we perceive the world, on what our interests are, on the uses we have for things, on the language we speak, and so on, it would be wrong for us to think that our classifications must correspond to objective features of the natural landscape. No doubt, we make the distinctions we do because we *interact* with the world in certain kinds of ways. But it is reasonable to believe that beings with other interests, or perceptual powers, or linguistic preferences, would interact with the world differently. Therefore, the human perspective is likely to be idiosyncratic. An objective, or natural, classification system which was not idiosyncratic would have to be based on *intrinsic* similarities or differences between the things that are classified together or distinguished from one another.

The elements of a natural classification system are natural properties and natural kinds. Things having the same natural properties are intrinsically the same in some respects. Things which belong to the same natural kinds are intrinsically the same in a number of respects. Plausibly, therefore, natural kinds are a species of property classes. And this, indeed, is the generally accepted view.

However, if natural kinds are supposed to be merely property classes, or the intersections of such classes, this would not explain their prominence in science as primary objects of investigation. On the generally accepted account, things are intrinsically the same or different from each other in various respects, according to the natural properties they happen to have. Things which have the same natural properties belong to a property class defined by these properties, and things which have different natural properties belong to different property classes. Thus, natural kinds, conceived of as property classes, supervene on natural properties.

It will be argued here, however, that property classes are not natural kinds. First, natural kinds belong to structured hierarchies,[7] whereas natural property classes do not. If a thing belongs to two different natural kinds, then one of these kinds must be a species of the other. But if a thing belongs to two different property classes, these two classes may not be so related. Secondly, a natural kind, unlike a property class, has no defining property. That is, there is no natural property which must, and can only, be possessed by things which are members of the kind. Certainly, there are natural properties which would distinguish, say, electrons from other kinds of things. Electrons must, for example, all have a certain mass, charge and spin. But there are very good reasons for saying that there is no property, such as that of *being an electron,* which all and only electrons can have.

As Armstrong clearly demonstrates,[8] there is an important distinction between properties and predicates. There are properties which have no names, e.g. because they have yet to be discovered, and there are predicates which truly apply to things, but which do not name properties. The question, therefore, arises: which predicates designate properties? I do not think that this question can be answered *a priori*. Nevertheless, there are some *a priori* principles which can be used to restrict the field of possible candidates. First, there is a *maxim of discovery*. If properties exist independently of language, then it cannot be the case that we can discover new properties just by the artful manipulation of language. New properties have to be discovered in nature, not inferred from the predicates we use to describe things. Secondly, there is a *principle of non-proliferation*: If something has a natural property P, then the only other natural properties which it *must* have in virtue of having P are just those on which P is ontologically dependent.[9]

Using this principle, in conjunction with the maxim of discovery, we argue[10] that Armstrong is basically right about disjunctive and negative properties,[11] but wrong about conjunctive properties. Conjunctive properties, even if the conjuncts are sometimes coinstantiated, may not exist. We also argue[12] that property names which are derived from natural kind names are at least suspect. Certainly, the property of being an electron (if, indeed, there were such a property) is not an ontologically fundamental one, and the same must be true of all properties (assuming that they exist) whose names are generated from natural kind names. Yet, electrons, protons, and the like, might well be ontologically fundamental kinds of objects. How is this seeming disparity to be explained?

I will assume, without argument, that there is a natural classification system involving natural kinds of objects, and natural properties. In particular, I will assume:

(1) that there are certain kinds of properties which exist objectively in nature, i.e. properties which exist independently of us, our interests, and our ways of conceptualizing the world. The causal powers, capacities and propensities of the most fundamental kinds of things are presumably properties of this kind. I call any such properties '*natural properties*'.

(2) that there are certain ontologically distinct kinds of objects existing in the world. These objects are naturally grouped into kinds, according to their intrinsic natural properties and structures. All such kinds of things I call '*natural kinds*'. The fundamental particles presumably all belong to natural kinds.

Natural kinds and natural properties are closely related. Indeed, it is often supposed that every natural property defines a natural kind, viz. the class of things having this property. Conversely, if K is a natural kind, then it is commonly supposed that there must be a corresponding natural property, viz. the property of being a K. For reasons already alluded to, and explained in detail elsewhere,[13] it seems that these inferences are both mistaken. Natural properties do not define natural kinds; and if K is a natural kind, then the property of being a K is bogus, or, at any rate, so different from other natural properties that it deserves to be classified separately from them.

Natural kinds are distinguished from other kinds of things by the following criteria:

1. The classes of things belonging to the various natural kinds are objective, and they belong, as classes, in a natural classification system. Call this the *objectivity requirement.*

2. Natural kinds are clearly distinct from each other. Where we are dealing with natural kinds, there are no borderline cases which call for arbitration. Call this the *distinctness requirement.*

3. Thirdly, all distinctions between natural kinds are based on intrinsic differences, not extrinsic ones. Things of the same natural kind are intrinsically the

same in some respects; things of different natural kinds are intrinsically different in some respects. Call this the *intrinsic identity requirement.*

4. Fourthly, natural kinds are distinguished from non-natural kinds by their associations with essential properties and real essences. Those intrinsic properties which determine the identity of a kind are its essential properties, and together these properties constitute the real essence of the kind. Let us call this the *essentiality requirement.*

5. Fifthly, if two members of a given natural kind differ essentially from each other, then they are members of different species of that kind. We call this the *speciation requirement.*

6. Finally, if anything is a member of two different natural kinds, then either one of these kinds is a species of the other, or there is a natural kind of which both of these kinds are species. I call this the *hierarchy requirement.*

Consider now the question whether natural properties define natural kinds. Natural properties are evidently like natural kinds with respect to the first four requirements. They are objective and mutually distinct, and their classes satisfy the intrinsicality and essentiality requirements on natural kinds. However, natural property classes do not satisfy the speciation or hierarchy requirements on natural kinds. For property classes do not speciate as natural kinds must. An electromagnetic wave and a sound wave may both have wavelength l. However, an electromagnetic wave and a sound wave are not the same kind of thing, even if they do have the same wavelength. They are not species of a natural kind whose members are things having wavelength l. For there is no such natural kind. Secondly, an object may have two distinct properties P_1 and P_2, neither of which is plausibly a species of the other, or of any more general property which includes both P_1 and P_2. An object might be both negatively charged and have a mass greater than four grams. But it is not clear that there is any general property which has these two disparate properties as species.

However, if natural property classes, i.e. classes defined by natural properties, are not natural kinds, then what are natural kinds? Are they a species of property classes, or are they something different altogether? Clearly, if the natural property which defines a given property class happens to be the conjunction of all of the essential properties of a certain natural kind K, then the class and the natural kind must have the same extension. For a thing can be a member of this class if and only if it is a thing of the kind K. However, there is no further property, over and above this conjunction of essential properties, which is the property of being a thing of the kind K. For a thing's being of the kind K is ontologically dependent on — indeed, it is ontologically reducible to — its having this conjunction of properties, not the other way around.[14] Therefore, by the principle of non-proliferation, we would need a separate argument for the existence of a property of being a member of a given natural kind.

If this is right, then it follows that there may be no such thing as the property of being an electron (or any other natural kind of thing). There is a natural kind which is instantiated by electrons, and, of course, there are natural properties, such as being negatively charged, which all electrons must have. There may be no property of being an electron which anything has or could have. Certainly, if there is such a property, then it is ontologically dependent on the essential properties of electrons. Therefore, the property of being an electron is not, and could not be, an ontologically fundamental property. However, electrons might well constitute a fundamental natural kind. Therefore, whichever way we look at it, there is an important difference between natural kinds and natural properties.

Perhaps the most important reason for thinking that natural kinds are special, and not to be assimilated to natural property classes, is this: natural kinds have an explanatory job to do. They signify certain clusters of causal powers, capacities and propensities which are naturally coinstantiated. There is every reason to think that these clusters exist independently in nature, and that they are ontologically fundamental. The fact that they are not classical universals need not worry us. With natural property classes, there is first of all the property; the class is then defined in terms of it. With natural kinds, there is first of all the kind; its essential properties have then to be discovered. To suppose that the class of electrons supervenes on the property of being an electron is to reverse the order of ontological priority.

At first, this result may seem counterintuitive. However, on reflection, it seems clear enough that the property of being negatively charged is different in a number of ways from the alleged property of being an electron. Many different kinds of things, including electrons, can have the property of being negatively charged. Clearly, nothing other than an electron can have the supposed property of being an electron. Even then, this is not really a property which the electron *has*, because to *have* this property it seems that it would have to *be* something else — a bare particular perhaps! I say that an electron *is* an electron; not something *which has the property of being* an electron. There is good reason, therefore, to conclude that there is no such property. There is no property of being a horse, either, and for the same kind of reason.

3. NATURAL CLASSIFICATION AND INTRINSICALITY

We all have a rough idea of what is meant by an intrinsic property or structure. It is a property that something has independently of any other thing. It is hard to say more precisely than this what an intrinsic property or structure is. Several philosophers have tried to do so, but their attempts have so far not succeeded, or not succeeded in explicating a concept that is relevant to the theory or practice of science. The reason for this lack of relevance is that the concept of intrinsicality they have sought is an *acausal* one. Hence, the independence required for intrinsicality has been assumed to be *logical independence*. What is needed for science, however, is a *causal* concept of intrinsicality, i.e. one which depends on a notion of *causal independence*.

For most philosophers, manifest shape is paradigmatically intrinsic. For the manifest shape of any object is logically independent of anything external to it. However, the manifest shape of an object is not necessarily the shape it has intrinsically, in the more interesting causal sense of this word. Let A be a stretched rubber band, and G its (manifest) shape. Now some object might have this shape intrinsically. A steel band, for example, might have this shape independently of the external forces acting upon it. It is not the case that if *any* object has this shape, then it necessarily has it intrinsically. If the rubber band were causally isolated from its surroundings, it would not have this shape. Intrinsicality in the causal sense is therefore not a property of properties, as many of those who have tried to explicate a logical concept of intrinsicality have supposed. This is because the very same property (having the stretched rubber band shape) may be an intrinsic property of one thing (e.g. a steel model of the stretched rubber band) and not an intrinsic property of another (viz. the stretched rubber band itself). Causal intrinsicality must therefore be either a relation between a property and its bearer which is distinct from the normal bearing relation, or else a property of that relation. For our immediate purposes, it does not matter which it is.

One philosopher who has tried to define a concept of intrinsicality is Jaegwon Kim.[15] Kim's definition of an internal or intrinsic property is equivalent to the following:

G is *intrinsic* $=_{df}$ possibly some object x has G at t, although no contingent object wholly distinct from x at t exists.

This definition of intrinsicality will not do for the causal notion, however, because it makes intrinsicality a property of properties. To capture the causal concept of intrinsicality, we might try the following:

An object x at t has G intrinsically $=_{df}$ x has G at t, and possibly no contingent object wholly distinct from x at t exists.

However, this will not do either. For G might be an intrinsic property of x even though x does not ever have G. A solid object, for example, has a certain intrinsic shape, viz. the shape it would have if it were not vibrating, rotating or distorted by external (differential) forces.[16] Unless the body happened to be in an inertial system in deep space, it would be distorted by gravitational or centrifugal forces and hence not have the shape which is intrinsic to it.

What is needed for science is a distinction between intrinsic and extrinsic properties which reflects the structure of causal explanations. Roughly, the distinction we need is between what is due to internal forces alone, and what is due, wholly or partly, to external forces or circumstances. This distinction is fundamental in the physical sciences, because, for example, it is needed to state the conservation laws. These laws all have the following form: Every event or process which can occur in nature is intrinsically X conservative, (where 'X' names the conserved quantity). The distinction is also needed for the construction of certain kinds of physical theories. The abstract theoretical models of physical systems are often designed precisely to represent the properties and structures of systems in causal isolation from their surroundings.

It would not be true to say that all theoretical models of physical systems are designed to represent the properties and structures which physical systems would have if they existed in causal isolation from their surroundings. It depends on what is to be explained by them. If what is to be explained is a system of relations which, let us say, does not depend at all on the intrinsic natures of the things related, then the appropriate theoretical model for this system of relations will reflect this fact. Our theories of space and spacetime, for example, have nothing to do with the intrinsic natures of the things that may be spatially or spatiotemporally related. Consequently, these things may be represented in our abstract theories by points or point-events—theoretical entities which have no intrinsic properties or structure, only location.

Other theoretical models are concerned with the patterns of interaction between two or more physical systems, and hence not only with the intrinsic properties of these systems considered in isolation. Nevertheless, these interacting systems may constitute a complex system whose elements are intrinsically disposed to interact in certain ways. A theoretical model of any such complex system will normally abstract from the various external forces which may act upon the whole complex, or on any of its parts, to focus upon the internal forces acting between the parts. Any external forces which may be acting on the system would be merely disturbing influences which are strictly irrelevant to the theory of the interaction. Consequently, they would, quite properly, be ignored in the theoretical model that is required to explain this interaction.

The intrinsic properties which are of most concern in the physical sciences are those properties which a thing has, and would continue to have, independently of the contingencies of its state or circumstances. They are not necessarily the manifest properties of anything, however, since the manifest properties of a thing are generally due to a mixture of causes, some internal and some external. The intrinsic properties of things might only be manifested, without the overlay of other effects, in idealized systems, in which they exist in causal isolation from all external forces.

It will be objected that to distinguish between intrinsic and extrinsic properties in this way leaves us in a kind of circle of interdefinable properties.[17] For the distinction we now want to make depends on that between internal forces or causes on the one hand, and external causal influences and relations on the other. How is this distinction to be made? The distinction cannot be made formally in the sort of way that Kim attempted to make his distinction between internal and external properties. For the distinction we have to make is not a logical distinction, but an ontological one that is to be made on the basis of the best scientific evidence available. For it is a distinction of fundamental *theoretical* importance, not one which can be made, as Kim tried to make his distinction, just by imagining the rest of the world away. The distinction has to be made in the light of the best currently available theories of the kinds of things under investigation, and of the causes of their behaviour. As our theories change, so might the intrinsic-extrinsic distinctions we should make change along with them.

Consider mass. It is now generally believed that the (rest) mass of an object is an intrinsic property of that object. That is, it is a property which the object would have, and would continue to have, if it existed as a closed and isolated system. It is not an *a priori* truth, however, that this should be so. If Mach had been right, then the dynamical properties of bodies, including their masses, would depend on the distribution of matter in the universe, and on their positions in relationship to it. In that case, mass would be an extrinsic property.

4. THE ESSENTIAL PROPERTIES OF NATURAL KINDS

Among the intrinsic properties and structures of things are those which make them the kinds of things they are. These properties and structures are their essential properties, and, together, they constitute the Lockean *real essences* of these kinds.

If the things in question are fundamental particles or fields, then their real essences will be just their causal powers, capacities and propensities. For these are the properties upon which the identities of such things depend. They have no shape or structure, or any other intrinsic categorical properties, upon which their identities could depend. They are defined by their causal roles.[18] If the natural kinds we are concerned with are more complex objects, such as kinds of atoms or molecules, then their identities will depend, not only on their more fundamental constituents, but also on their structures. Things of the same natural kind will be constituted by things of the same kinds, interrelated in the same ways.

If the things we are dealing with are natural kinds of causal processes, then these will, necessarily, be displays, appropriate to the circumstances, of the same natural dispositions. One might, for example, be able to fake refraction, perhaps by using a system of mirrors and a diffraction grating, but if the process is not a display of the refractivity of some material, then it is not refraction.

Thus, there is an intimate relation between the real essences of natural kinds of processes and the natural powers, capacities and propensities of things, since the latter are the displays of the former. Hence, knowledge of the causal powers, capacities and propensities of things gives us, *eo ipso,* knowledge of the natural processes in which they may participate. Hence, if we know the causal powers, capacities and propensities of the kinds of objects involved in a process, and how these objects are arranged or constituted, then we must be able to explain, and to predict, how they will (or with what probability they would) act or interact.

Inevitably, therefore, one of the basic aims of physical science must be to discover what natural kinds of things exist, and what their causal powers, capacities and propensities are. Indeed, I would suppose this to be the primary aim of physical natural science. At its most fundamental level, science is just the search for essence.

5. AN ONTOLOGY OF NATURAL KINDS

Since natural kinds are hierarchically related by the species relationship, we may think of a family of natural kinds as a tree-like structure, with branches representing species. The most general kinds in the family will be represented by the lowest branches, and the most specific kinds by the highest branches. In the case of a spectral kind, the branches will cluster into a dense fan, each radius of which will be its own unique branch, and will represent a species of the spectral kind which is distinct from every other species of that kind.

The most specific members of any family of natural kinds, or of any spectral kind, must, by nature, be intrinsically identical with respect to their causal powers, capacities and propensities. For if there were any natural intrinsic differences in any of these respects between any two members of a species, then they would have to be members of different natural kinds, and hence members of distinct subspecies of those species. All electrons, for example, would appear to be intrinsically identical. But if some electrons were found to have an intrinsic property which others lacked, then we should have to acknowledge that there were different species of electrons. Let us call the most specific kinds which are species of any given natural kind the *ultimate species of that kind,* and the ultimate species of natural kinds generally the *ultimate species.*[19]

As the most specific members of a family of natural kinds or spectral kinds must be intrinsically identical in their natural causal powers, capacities and propensities, it follows that the types of processes generated by these powers or propensities must also be intrinsically the same. For natural processes are just the expressions of such natural dispositional properties as these. The natural processes which occur will, of course, belong to natural kinds, and, as with natural kinds of properties, natural kinds of processes will be hierarchically related by the species relationship. The rusting of iron, for example, is a species of oxidation, which is a species of chemical combination, which is a species of chemical interaction, which is a kind of physical process.

The theory of natural kinds, as it applies to processes, thus leads to a natural hierarchical classification of processes, with some more fundamental than others. Given a traditional ontology of universals and particulars, however, no such natural hierarchical classification of processes is to be expected. Indeed, since individual processes do not ordinarily occur in isolation, but do so along with other processes occurring to the same things at the same time, individual processes are not classical particulars. Since they are manifestly not universals, they would appear to have no place at all in the traditional ontology.

Given that there are natural kinds of entities and processes, and that the members of any ultimate species must be intrinsically identical to each other in certain basic respects, it is reasonable to speculate that there is a *universal* which they all instantiate. Let us suppose that this is so, and call such a universal a *natural kind universal.*

Clearly, there would have to be different sorts of natural kind universals, depending on the kinds of ultimate species they refer to. If the members of an ultimate species are physical objects, then the universal will be instantiated by such objects. If an ultimate species is a kind of process, then the processes of this kind must be intrinsically identical, and therefore the universal would have to be one which has ultimately specific processes as its instances. If an ultimate species is a species of properties, e.g. of specific quantitative properties or relations, then the universal would have to have such properties or relations as its instances.

All of these supposed natural kind universals would stand in contrast with the kinds of universals usually discussed in the literature. For these are associated with natural properties, rather than with natural kinds. Let us call these ordinary universals *ordinary universals.*

The crucial difference between natural kind universals and ordinary universals would be this: since there is no property of being a member of a natural kind, the supposed natural kind universals could not be instantiated *in* things in the way that ordinary universals are. Natural kind universals could only be instantiated *by* things. If it were a natural kind universal which ranged over physical objects, then it could only be instantiated by physical objects. If it were one which ranged over properties, then it could only be instantiated by properties, and so on. Ordinary universals, on the other hand, can be instantiated *in* things. For example, the cloud overhead might be negatively charged. But if so, the ordinary universal *negative charge* would be instantiated *in* the cloud, not *by* it. The cloud itself is obviously not an instance of the property of being negatively charged.

Natural properties can be possessed by many different kinds of things, both natural and non-natural. Therefore, most ordinary universals would have to be similarly promiscuous. Moreover, the same property can be an essential property of one kind of thing, but only an accidental property of another. For example, while the property of being negatively charged is an essential property of both the electron and the anti-proton, it is only an accidental property of a negatively charged raindrop. Ordinary universals are therefore not tied to natural kinds as natural kind universals would obviously have to be.

However, natural kind universals are in some ways unlike ordinary universals. First, the most fundamental kinds of properties in nature would all appear to be quantitative. That is, they come in degrees. Therefore, there is strong evidence that quantities, rather than specific quantitative properties or relations, are the truly fundamental entities. If this is right, then the fundamental quantities, which are natural kinds of specific quantitative properties and relations, are ontologically more fundamental than any of their instances. Similarly, it might be supposed that the natural kinds of particles and fields are ontologically more fundamental than the members of these kinds, and that the natural kinds of processes that occur in nature are ontologically more basic than any instances of these processes. If so, then the theory of natural kinds leads to something like a Platonic theory of forms, rather than an Aristotelian theory of universals. However, I do not yet wish to be committed to such a theory.

Whatever position we take on the ontological status of natural kinds, it is clear that they are fundamental entities which are different from Aristotelian universals. First, natural kinds belong in families, and to specify a natural kind, we must not only say what its essential properties are, but what species, if any, it includes, and to what family it belongs. Secondly, the members of a natural kind need not all be intrinsically identical. For spectral kinds, at least, there are ranges of permitted intrinsic variation. Thus the various instances of a quantity may differ from each other, as may the instances of a given natural kind of process. Hence, to specify such a kind, we must state the range, or ranges, of intrinsic variation it permits.

6. THE LOGIC OF NATURAL KINDS

Assuming that one accepts a natural kinds ontology, the question arises how one should proceed to do natural science, and what inferences one should consider to be justified. Obviously, this is far too big a question to be answered properly here. However, there are some things we can say at once about the logic of natural kinds.

First, it follows from what has already been said informally about natural kinds that the following theses are *necessarily true*:

(1) If $x =_i y$ then $x =_e y$.

(2) For every K, there is an intrinsic property P such that PeK.

(3) If $x \in$ K and PeK, then $\Box Px$.

(4) If PeK, and $K_1 \subset K$, then PeK_1.

(5) If $x \in K_1$ and $K_1 \subset K_2$, then $x \in K_2$.

(6) If $x, y \in K$, and $x =_i y$, and there is a K_1 and K_2 such that $x \in K_1$, $y \in K_2$, $K_1, K_2 \subset K$, and $K_1 \neq K_2$.

(7) If $K_1 \neq K_2$, then there is a property P such that Pe$K_1 \not\equiv$ PeK_2.

(8) If $K_1 \subset K_2$, and $K_2 \subset K_3$, then $K_1 \subset K_3$.

(9) If $x \in K_1, K_2$, and $K_1 \neq K_2$, then either $K_1 \subset K_2$, or $K_2 \subset K_1$, or there is a K such that $K_1, K_2 \subset K$.

Where, '$x \in K$' = 'x is a member of the natural kind K'.
'PeK' = 'P is an essential property of K'.
'$K_1 \subset K_2$' = 'K_1 is a species of K_2'.
'$x =_e y$' = 'x is essentially the same as y'.
'$x =_i y$' = 'x and y are intrinsically identical in their causal powers, capacities and propensities'.

To these we may now add the following:
(10) For all x, ($\Box x \in K$ or $\Box x \notin K$).
(11) There are no two natural kinds, K_1 and K_2, such that *necessarily*, for all x, $x \in K_1$ or $x \in K_2$.
(12) The class of things defined as the intersection of the extensions of two distinct natural kinds K_1 and K_2 is not necessarily the extension of a natural kind, unless $K_1 \subset K_2$, or $K_2 \subset K_1$.
(13) The class of things defined as the union of two distinct natural kinds K_1 and K_2 is not necessarily the extension of a natural kind, unless $K_1 \subset K_2$, or $K_2 \subset K_1$.

What (10) says is that the membership of natural kinds is never a contingent matter. If x is a member of a natural kind K, then it is so necessarily. If it is not a member, then this is not just accidental. It could not be or become a member. (11) affirms that if K_1 is a natural kind, then there is no complementary natural kind K_2 (= not K_1), although the class of things which are not K_1 might *accidentally* be the extension of a different, and perhaps wholly unrelated, natural kind K_2. That is, there are no negative natural kinds. If the class of ravens is a natural kind, then the class of non-ravens is not. In this respect, natural kinds are like natural objects. If Brian Ellis is a natural object, then non-Brian Ellis is not also a natural object (although, of course, it could be the case that there is only one other natural object in the universe). (12) and (13) assert that the intersections and unions of natural kind classes need not also be natural kind classes, unless one of these classes is a species of the other.

In all of these principles, the kind of necessity involved is metaphysical. What is necessary in this sense is, like what is logically necessary, true in all possible worlds. However, what is metaphysically necessary is not *a priori*. That is, it has to be discovered by empirical investigation.

7. REASONING ABOUT NATURAL KINDS

Given a natural kinds ontology, a primary aim of physical natural science must be to discover what natural kinds of substances, properties and processes exist, and what their essential natures are. The big problem, then, is how we are to achieve these objectives.

Probably, the history of chemistry illustrates, as well as the history of any science could, the kinds of investigative processes by which these aims may be realized. For we can be fairly confident that chemists now know many of the distinct kinds of chemical substances that exist naturally on Earth, or can be made in laboratories, how these substances are interrelated, what makes each of them the kind of substance it is, and how and why it acts, or interacts with other substances in the kinds of ways it does. This has been achieved by an investigative process which has involved the parallel development of a hierarchical chemical classification system (of elements and compounds), a theory of chemical identity adequate for this system (the atomic-molecular theory), an extensive knowledge of the chemical properties of the various kinds of substances so classified, and sophisticated theories to explain the sources of these properties, and the kinds of chemical processes for which they are responsible.

The process by which all of this chemical knowledge and understanding was achieved is very complex, and it involved many different kinds of investigations. It required the isolation and purification of the different kinds of substances, and the systematic study of their manifest and dispositional properties. It needed the development of theories of chemical processes, adequate to explain the various kinds of chemical changes and reactions which can seen to occur, compatibly with the underlying theory of chemical identity. It required the discovery and explanation of the various kinds of chemical properties which the different substances can have.

The kinds of investigations involved in this complex process are all strongly interdependent. Hypotheses about what kinds of substances there are may depend on theories of chemical properties or processes, and conversely. Lavoisier's theory of acidity, for example, led him to believe that there must be an elementary substance which, when oxidized, produces spirit of salt. For Lavoisier believed that acids were essentially oxygenated bases or radicals. Thus we had carbonic acid (CO_2), sulphurous acid (SO_2), nitrous acid (NO), nitric acid (NO_2), and so on. Lavoisier argued, therefore, that what we now know as hydrochloric acid must also be an oxide. He called the postulated base of this oxide 'muriatic base', and the acid 'muriatic acid', but lamented that 'this acidifiable base adheres so very intimately with oxygen, that no method has hitherto been devised for separating them'.[20]

As complex as this process of reasoning may be, it is clear that it involves natural kind reasoning about kinds of substances, kinds of properties and kinds of processes, and the development of theories concerning the essential natures of all of these natural kinds. Therefore, in the field of chemistry, at least, this kind of reasoning must be regarded as fundamental. The process of acquiring chemical knowledge cannot possibly be understood, if we do not know the nature of such reasoning.

All natural kind reasoning seems to depend on hypotheses about what natural kinds of objects, substances, properties and processes exist, and theories about what their essential natures are. If we believe that K is a natural kind of substance, for example, and that *x* is an instance of this kind, then we will suppose that the intrinsic causal powers, capacities and propensities of *x*, and, if it is a complex substance, then its intrinsic structure as well, are properties it has necessarily, in common with all other instances of this kind. For these intrinsic properties, and this structure, will be immediately identified as the essential properties and structure of the kind. Therefore, once the kind is known, the empirical task is to determine what the intrinsic dispositional properties and the intrinsic structure of any particular instance of the kind are.

Natural kind reasoning about kinds of properties and processes is similar. It proceeds from assumptions about the kinds of properties and processes that exist, and what kinds of properties or processes are instantiated in the particular case, or cases, being investigated.[21] There are, however, some important differences between natural kinds of substances, on the one hand, and natural kinds of properties and processes, on the other. First, natural kinds of properties and processes are typically spectral. That is, their instances are not all identical, *qua* properties or processes, but differ from each other quantitatively in one or more ways. Consequently, the essences of these kinds must be described quantitatively — as the measures of their influences in causal processes, in the case of causal powers and capacities, and as quantitative relations between causes and effects, in the case of causal processes.

Secondly, properties and processes are often hidden, or their effects swamped by other properties and processes. Indeed, some properties and processes rarely, if ever, occur in isolation, and often their effects cannot be measured directly. Consequently, to describe the effects of such properties or processes, it is often necessary to abstract from anything that can actually be observed to consider what would happen in the imagined absence of other factors that exist in the actual situation. As a result, descriptions of the essences of causal properties and processes are often abstract, and expressed either categorically, as statements about the behaviour of idealized objects in ideal circumstances, or subjunctively, as conditionals about how real objects would behave, if they, and the circumstances of their existence, were ideal.

Although there are these differences between natural kind reasoning about kinds of properties and processes, on the one hand, and substances, on the other, the basic structure of the reasoning is the same.[22] Any instance of a natural kind of property or process is, in principle, as good as any other instance as a basis for generalizing to all members of the kind. Indeed, it serves as a basis for the claim that all members of the kind *must* have certain effects, (depending on the magnitude of the cause), or *must* proceed in such a such a way, and have such and such effects, (discounting the influence of other causes). That is, natural kind reasoning leads, not only to universal generalizations about substances, properties and processes, it implies that these universal generalizations are, in some sense, necessary.

The statements which describe the essences of the natural kinds of causal processes are a species of laws of nature. They are the laws we call 'causal laws'. The causal laws are therefore necessary in the sense in which any statement ascribing an essential property to a natural kind is necessary. That is, they are metaphysically necessary. The theory of natural kind reasoning thus explains, not only the process by which we may arrive at a knowledge of causal laws. It also explains the modal status of these laws.

La Trobe University, Bundoora, Victoria

NOTES

* This essay arises out of work done in collaboration with Caroline Lierse, and owes much to her.

1 See Ellis and Lierse (forthcoming), hereafter referred to as E & L, for a more complete development and justification of the ontology of natural kinds.

2 E & L, Sections 2-5.

3 E & L, Section 8.

4 E & L, Sections 2-5.

5 E & L, Section 8.

6 See E & L, Sections 2-5, where it is argued that there is a clear distinction to be made between natural properties and relations, on the one hand, and natural kinds on the other.

7 E & L, Section 6.

8 Armstrong (1978) and (1989).

9 For an analysis of these two principles, and of their implications for the ontology of properties, see E & L, Sections 2-5. The concept of ontological dependence required for the principle of non-proliferation is developed in Section 5.

10 E & L, Sections 2-5.

11 For reasons, which are independent of those given here, Armstrong ((1989), pp. 82-84) argues that there are no disjunctive or negative universals. However, he has no objection to conjunctive universals ((1989), p. 84).

12 In E & L, Section 4.

13 E & L, Sections 2-5.

14 E & L, Section 5.

15 See Kim (1982), and Lewis's (1983) discussion of Kim's paper.

16 The concept of a differential force is due to Hans Reichenbach. Reichenbach used this concept to define a rigid body:- 'Rigid bodies are solid bodies which are not affected by differential forces, or concerning which the influence of differential forces has been eliminated by corrections; universal forces are disregarded.' (Reichenbach (1957), p.22). Thus, as defined by Reichenbach, a rigid body is a theoretical ideal. It is simply any body, ideal or actual, whose shape is just its intrinsic shape.

17 Lewis (1983), p.197.

18 I use the word 'define' to signify real definition, as opposed to mere nominal definition.

19 Classically, these were called 'species infima'. However, since this term is not widely used these days, and we wish to dissociate ourselves from the Aristotelian theory of their ontological priority, we shall not use this term here.

20 Lavoisier (1949), p.70.

[21] Several philosophers have argued recently that natural kind reasoning concerning properties and processes involves consiliences of inductions, allowing what appear to be different kinds of processes to be seen as processes of the same kind. (Butts (1977); Forster (1988); Harper (1989)). I would agree that such consiliences of inductions do occur, and often signal important advances in conceptualization. However, I do not think they describe kind of reasoning that is characteristic of subjects like Chemistry. I think Macnamara (1991) is right about that.

[22] It is accurately described in Macnamara (1991), in so far as the reasoning is concerned with natural kinds of substances. I think exactly the same kind of reasoning is apt for natural kinds of properties and processes.

REFERENCES

Armstrong, D.M. (1978): *Universals and Scientific Realism* (2 vols). Cambridge, Cambridge University Press.

— (1989): *Universals: An Opinionated Introduction.* Boulder, San Francisco, London; Westview Press.

Butts, R. E. (1977): 'Consilience of Inductions and the Problem of Conceptual Change in Science' in Colodny, R. G. (ed), *Logic, Laws and Life.* Pittsburgh, University of Pittsburgh Press, pp. 71-88.

Ellis, B. D. and Lierse, C. (1994): 'Dispositional Essentialism', *Australasian Journal of Philosophy* **72**, pp 27-45.

Ellis, B. D. and Lierse, C. (forthcoming): 'The Fundamental Importance of Natural Kinds'. Paper read to Australasian Association of Philosophy Conference, Adelaide, July, 1993. Available as *Victorian Centre for the History and Philosophy of Science Preprint, 3/1994.*

Forster, M. R. (1988): 'Unification, Explanation and the Composition of Causes in Newtonian Mechanics', *Studies in the History and Philosophy of Science* **19**, pp. 55-101.

Harper, W (1989): 'Consilience and Natural Kind Reasoning', in J. R. Brown and J. Mittelstrass (eds), *An Intimate Relation.* Kluwer Academic Publishers, pp. 115-52.

Kim, J. (1982): 'Psychological Supervenience', *Philosophical Studies* **41**, pp. 51-70.

Lavoisier, A. (1949): *Elements of Chemistry,* Book 1. Chicago, Illinois, Henry Regnery Company, Publishers for the Great Books Foundation.

Lewis, D. K. (1983): 'Extrinsic Properties', *Philosophical Studies* **44**, pp. 197-200.

Macnamara, J (1991): 'Understanding Induction', *British Journal for the Philosophy of Science* **42**, pp 21-48.

Reichenbach, H. (1927): *Philosophie der Raum-Zeit Lehre,* tr. by M. Reichenbach and J. Freund as *The Philosophy of Space and Time* (1958). New York, Dover.

CAROLINE LIERSE

THE JERRYBUILT HOUSE OF HUMEANISM *

1. INTRODUCTION

The ontology of analytic metaphysics consists primarily of fundamental properties and relations. This ontology is variously enriched with a variety of metaphysical apparatus, ranging from supervenient relations, possible world accessibility relations, through to primitive causal powers that determine physical behaviour. The role of such metaphysical additions is to augment the ontology to yield a fully fledged metaphysics that can explain the nature of the world.

It is from this common basis and with this common purpose that philosophical battles rage over what supplementary metaphysical machinery is required to furnish a complete and consistent metaphysics. With few exceptions, however, the battleground where such conflicts occur has a distinctly *Humean* landscape. For the properties and relations that form the ontological terrain all conform to Hume's Distinctness Thesis, *viz*, the thesis that there can be no necessary connections between distinct existences.[1]

Nonetheless, in recent times new metaphysical programmes with a distinctly *un*-Humean ontology have emerged.[2] But by and large, this anti-Humean sentiment is only embraced by a minority. The central reason for this state of affairs lies in the fact that Humeans see little reason to convert. True, they will admit that anti-Humean theories may have *relative* advantages when it comes to analysing notions such as dispositions and nomic necessity. But, they argue, such benefits come at a cost, and the cost is the violation of the Distinctness Thesis — a thesis they see no sound reason to abandon.

It usually is held that the onus of proof on why it should be abandoned lies with those who choose to reject it. Given the *prima facie* plausibility of the Distinctness Thesis, I accept this as a fair challenge. However, in the sections that follow I will not proffer a refutation of Hume's central metaphysical maxim. For I am not explicitly seeking to persuade those who genuflect before Hume's alter to abandon the Distinctness Thesis. Rather, I simply seek to demonstrate that they have *already* abandoned it. Thus the criticisms of *anti*-Humean theories that Humeans are apt to offer cannot be couched in terms of a violation of the Humean maxim. For given that the Humeans *themselves* systematically violate it, the onus now rests on them to modify their theories to remove the offending inconsistencies, or else explain why the Distinctness Thesis should not be abandoned completely.

P.J. Riggs (ed.), Natural Kinds, Laws of Nature and Scientific Methodology, 29–48.

2. THE CHARGE

Humeans use the Distinctness Thesis as a metaphysical principle for a number of diverse applications.[3] However, its most widely used and acknowledged application is with respect to physical properties, and the sorts of relations that connect such properties. Given that the Distinctness Thesis prescribes that there can be no *necessary* connections between physical properties, this entails that all relations between such properties, *including* those relations which seem to manifest themselves with exceptionless regularity — such as the laws of nature — are contingent. For according to the Distinctness Thesis, there is no reason why we *necessarily* have the laws that we do. It is perfectly within the realms of possibility that different relations, and hence different laws could link the different kinds of physical properties. Thus conceived, the Distinctness Thesis entails a contingency theory of laws.[4]

Given that the Distinctness Thesis permits the possibility of the same kinds of properties being governed by different sets of laws, it follows that there is no *necessary* connection between what a thing is, and how it is disposed to behave. Thus the laws of nature that describe a property's disposition to behave have no necessary connection to the property to which the behaviour is ascribed.

This lack of a necessary connection between the intrinsic character of a property, and the behaviour it exhibits, is deeply entrenched in the current metaphysical framework. On the surface, this logical wedge seems entirely plausible. Different objects each display a vast array of behaviours, and exhibit no apparent connection between the type of object they are, and the type of behaviour they display. For instance, drop a squash ball and it will most probably bounce, however the same is not true for a porcelain vase. If we were to drop such a vase, it would most likely shatter. But our expectations about these respective outcomes do not result from perceiving a *necessary connection* between the kind of object and the behaviour it displays; rather, our expectations are driven by inductive inferences based on the outcomes of past experiences.

Hence porcelain vases break when (suitably) dropped, the Humean argues, *not* because there is any necessary connection between the property of being a porcelain vase and "breaking" behaviour, but because that is the way things happen to be *in our world.* It is *logically* consistent to imagine porcelain vases that bounce, and hence the connection between object and behaviour is a contingent one. This contingency, the Humean argues, also explains why science is an *a posteriori* enterprise. For given that the connection between object and behaviour is contingent, there is no *a priori* way of matching different types of objects with different kinds of behaviour.

One salient consequence of the Distinctness thesis is that it entails that the intrinsic nature of physical properties must be specified *without appealing to the kind of behaviour they tend to exhibit.* The nature of a property *cannot* be specified in terms of how it causally interacts and thus in terms of the behaviour it is disposed to manifest, since this state of affairs is determined by the laws of nature, which have only a contingent relationship with the property.

However, the upshot of this restriction that the Distinctness Thesis imposes is that it makes ontological discourse extremely difficult and sets metaphysics adrift from its empirical moorings. For as Humeans insist on divorcing an entity's identity from its behaviour, *no* causal interaction *nor* any aspect of its behaviour can provide an insight into the property's true nature. In fact, it is perfectly consistent with Humean metaphysics for two properties with intrinsically *different* natures to exhibit the same behaviour, and hence be nomically indistinguishable. The consequence of this is that the Humean is left with an ontology consisting of properties whose natures are *ex hypothesi* inscrutable.

An adherence to the Distinctness Thesis also poses a further difficulty, namely, how do Humeans determine what to admit into their ontology of *physical* properties? Possessing causal efficacy[5] is an obvious criterion for physicality, but according to Humeanism, the causal role an entity plays is dependent on the laws of nature, which are said to be contingent. Therefore, it follows that an object's causal efficacy is only a contingent property of an object — a property it possesses in some worlds and not others. But if Humeans wish to divorce the nature of a property from its behaviour *without* making physicality a world - relative property, then a criterion for physicality cannot be furnished in terms of an appeal to causal efficacy. So how then do Humeans determine what properties are physical? The two problems that the Humean must confront can be formulated thus:

(i) What is a *Humean* criterion for physical/ontological existence?

(ii) How does a Humean distinguish one type of physical property from another?

Answers to these two questions are necessary preliminaries that need to be addressed before any empirically meaningful philosophy can proceed. This is not to say that it is necessary that all Humeans be in agreement with respect to the individual responses they provide. We may argue about the relative merits of the answers to these questions, but the point is that some response must be given.

However, I argue that the only reasonable answers available necessitate violating the central principle of Humeanism, *viz*, the Distinctness Thesis. For if an ontology of inscrutable matter is to be avoided, then reference must be made to the causal powers that the properties possess. Furthermore, if a property's causal powers are to have any relevance to the identity of the property, then the connection between the nature of the property and the kind of behaviour it displays must be more than merely contingent. Hence I suggest the following *un-Humean* criteria:

(i) *Causal Criterion for Physical Existence*: A necessary condition[6] for a property to gain admission into our (physical) ontology is that it be causally efficacious in some respect.

(ii) *Causal Criterion for Qualitative Identity*: The qualitative character of a (physical) property is determined fundamentally by the kind of behaviour it is disposed to manifest. That is, two properties that are disposed to display the same behaviour in the same set of circumstances are qualitatively identical.

The first criterion is primarily an epistemic one. Admittedly, there is little to justify it other than Occam's Razor. For it seems that for us to be capable of having knowledge of (i.e., be aware of) a physical property it is a necessary condition that the property can causally interact with us directly, or else by way of a causal chain. We are, of course, open to reject this criterion and admit causally idle properties. For instance, causally inefficacious pink piglets that fly around the rings of Saturn qualify as causally idle properties. But relaxing this condition would grossly inflate our ontology. Moreover, how would *a posteriori* knowledge of such properties be possible?

However, it should be noted that subscribing to this criterion does not *in itself* constitute a violation of the Distinctness Thesis. This is because it only states that physical properties must be causally efficacious. It does not prescribe that there is a necessary connection between a *particular* property and its *particular* causal powers. It merely states that to *physically* exist, a property must have at least one causal power — any kind will do.[7] This still leaves the property quite free to have a different causal power in different possible worlds; consequently, the relation between the property and the *particular* causal power it possesses maintains its contingent status.

The second criterion has a similar epistemic flavour to the first. This stems from the fact that the two criteria are not unrelated, for the justification of the second criterion is partly dependent on the justification of the first. Given that we can only know of a property through its causal interactions, it follows that we can only determine whether two properties are of the same physical type by comparing types of behaviour. Thus properties that are disposed to instantiate identical behaviour in identical circumstances are deemed to be qualitatively identical.

However, subscribing to this criterion *does* necessitate a violation of the Distinctness Thesis, for in this case there *is* a direct link between the qualitative character of a property, and the causal powers it possesses. In this instance, the connection is *not* contingent.

I argued earlier that adherence to the Distinctness Thesis *ex hypothesi* entails an ontology of inscrutable substance. However, I know of no current Humeans who countenance this as part of their ontology. So, given that Humeans need to supply a set of criteria, and the set just proffered was distinctly *un-Humean*, the question that must be asked is "what set of criteria do Humeans embrace?"

I predict that very few Humeans would wish to irrevocably divorce themselves from empiricism. An attentive examination of their philosophical papers vindicates this prediction. For the unhappy fact is that they do abandon the Distinctness Thesis and embrace a causal criteria in one form or another. However, in most papers of a Humean ilk, this prohibited metaphysical baggage is never explicitly declared. It is usually concealed in arguments, in the form of unstated premisses. Nevertheless, with careful reading it can be seen that this metaphysical contraband infiltrates the whole Humean philosophical edifice.

However, I am not suggesting that the Humeans are deliberately concealing a metaphysical flaw. The fact that their subscription to a causal criterion almost always is unacknowledged most probably stems from the fact that philosophy usually is done in a piecemeal fashion, where the fundamentals are more or less assumed to be in place. Hence, the Humean's illicit espousal of a causal criterion is unlikely to surface in isolated philosophical papers. Instead, we are more likely to find *explicit* evidence in philosophical writings that are constitutive of a comprehensive Humean research programme. Of the current programmes in existence, the most well known and fully articulated are those developed by David Armstrong and David Lewis.

3. THE EVIDENCE

Although David Armstrong would not consider himself a Humean,[8] he does embrace Hume's Distinctness Thesis, so for present purposes, he qualifies to be granted the name. He describes himself as a scientific realist, and accordingly, embraces a sparse ontology of properties of whose existence must be determined *a posteriori*. His properties take the form of universals, thus he is an *a posteriori* realist. His formal criterion for admission into his ontology is as follows:

> [W]e may appeal to a plausible necessary condition for something's being a property. If a particular has a property that property must endow the particular with some specific causal power, or if the property is causally idle then it must at least be an intelligible hypothesis that the property should endow particulars with some specific causal power.[9]

So far, things don't look too good. Although Armstrong has not made a *direct* connection between individual properties and *kinds* of causal powers they possess, the seeds of this union are already planted.

Given that he has provided an existence criterion, what is now needed is a criterion for *a posteriori* distinguishing one *type* of property from other. This is a two step procedure. Firstly, a conceptual analysis is needed to define the notion of qualitative identity. With respect to this first step, the kind of analysis offered is dependent on the kind of ontology embraced. But as Armstrong's ontology consists of universals, this step is relatively straightforward. In this case, universals can be used to analyse qualitative identity. In an Armstrongian ontology, two properties are qualitatively identical if and only if they instantiate the same universal.

Now for the second step — the *a posteriori* criterion for type identity. Armstrong's criterion is as follows:

> Formal identity criteria for universals may be given. They are identical if and only if they bestow identical causal powers upon the particulars which fall under them.[10]

And here we have it — in this instance, an out-right violation of the Distinctness Thesis. Armstrong explicitly defines sameness of type in terms of sameness of causal power. Thus the *contingent* connection between a property and its behaviour has been abandoned.

However, regrettably for Armstrong, the rejection of the Distinctness Thesis at this critical point entails that he must also reject his contingency theory of laws, and make the appropriate modifications to his combinatorial theory of possibility. For by identifying the nature of properties with their causal powers in this way, it is not logically consistent for Armstrong to assert that the laws governing these properties could be other than they are. Thus in all possible worlds, if a particular property exists, then it possesses its kind of causal power *necessarily*, for the simple reason that the kind of causal power it possesses *determines* the type of property it is.

Paradoxically, what Armstrong fails to recognise and yet explicitly endorses is the fact that there *is* a connection between the nature (i.e., "geography")[11] of a property and the behaviour the property displays. For instance, given our understanding of the concept of negative charge, a particle could not be said to possess the property of negative charge and fail to manifest the behaviour of repelling like charges. This connection between charge and "charge-like" behaviour is analytic: having negative charge necessarily entails having the propensity[12] to repel other things possessing negative charge. Hence, the nature of a physical property is inescapably conceptually intertwined with its causal and nomological role.[13]

Another prominent Humean research programme has been developed by David Lewis.[14] The Lewisian metaphysics provides an interesting contrast to Armstrong's, for it differs fundamentally in two major respects. Firstly, and most significantly, Lewis does not want to be (inescapably) committed to the existence of universals, and thus he constructs his ontology from a nominalistic theory of properties. And secondly, the members of his ontology are admitted on *a priori* grounds. Causally inefficacious properties are not automatically prohibited from being granted membership.

Although Lewis's metaphysics is an *a priori* metaphysics *par excellence,* this does not negate the fact that if his philosophical programme is to have empirical relevance, it must supply a criterion for distinguishing the *physical* properties from the abundantly many other properties that populate his ontology. Secondly, it must supply a criterion for determining qualitative identity with respect to these physical properties.

As far as I can tell, Lewis offers no formal conceptual analysis of physicality, preferring instead to defer to physics.[15] However, such deference to physics is a metaphysically dangerous policy. For by handing over the responsibility of determining the conceptual criteria for physicality to the physicists, Lewis thereby abdicates control of such criteria and thus lies prone to whatever metaphysical standards physics ultimately espouses. If such standards are friendly to his philosophical programme, well and good. But unfortunately for Lewis, this does not seem to be the case. For the current metaphysical presuppositions underlying physics rely heavily, *if not exclusively*, on a causal criterion for physicality. Broadly speaking, from a physicists' point of view, for a property to qualify for admission into the ontological arena of physics, it must satisfy the following two regulative principles:[16]

(1) *The Principle of Multiple Manifestation*: The principle that a real property must manifest itself in more than one way (i.e., it must have detectable consequences, in other words, be causally efficacious).

And

(2) *Schlesinger's Principle of Connectivity*: The requirement that any real property must be nomically connected with other properties, so that two physical systems cannot differ only in respect of a single property.

The first principle is explicitly a *causal* one. Thus if Lewis is to defer to the authority of physics to supply the conceptual analysis, he is engaging in a potentially dangerous liaison with a causal criterion. Of course, it could be the case that in the final scheme of things, physics may abandon one or both of these principles. However, at this point in time, I know of no plausible way that physics could determine its inventory of physical properties other than by adhering to these constraints. Unless Lewis can offer a conceptual criterion for physicality which is friendly to physics but does *not* invoke a causal criterion, he is already drifting into distinctly un-Humean territory.

The second requirement of Lewis's metaphysics is an analysis of qualitative identity. As we saw earlier, this is a two step process. Firstly, Lewis needs to supply a conceptual analysis of qualitative identity, and secondly, he needs a criterion for empirically distinguishing one type of property from another.

As stated previously, Armstrong's conceptual analysis of type identity is relatively straightforward. In his realist metaphysics, qualitative identity is almost taken as a primitive in that it is defined in terms of participation in the same universal(s). However, as Lewis does not want to commit himself to an ontology of universals, this two step analysis must be fashioned in a manner that does not require their services.

Lewis begins with nominalistic theory of properties. However, his ontological menu is an extraordinary rich and generous bill of fare. Any logically consistent property or complex property, no matter how bizarre, or ineffectual, is admissible. If gruesome disjunctive gerrymanders are your thing, these can be served to order. But for an ontology to be saleable in the metaphysical marketplace, it must be able to earn its keep. And one necessary prerequisite is that it can provide an analysis of qualitative identity.

However, Lewis's population of properties is an unruly bunch. As his ontology is so generous, it is also undiscriminating. Sameness of type, as normally understood, cannot be analysed in terms of sameness of properties. Two things, which we take to be of the same type, may both share the properties we think salient to type identity, but differ in an infinite number of other irrelevant ones — they may have different spatio-temporal positions, for instance. But as things stand, there is no way to mark out the relevant ones. There are simply too many properties, no clear job allocations, and not enough meaningful work to go around. So before Lewis can establish type identity, he needs some criteria for partitioning off the irrelevant free-loaders from the properties capable of doing the required metaphysical work.

Sameness of nature, Lewis argues, is related to the concept of *duplication*.[17] Perfect duplicates, it seems, are qualitatively alike. But, as stated earlier, "qualitatively alike" does not translate to "sameness of properties" in an unrestricted sense, so this alone will not do. Obviously a more refined definition is needed. What we mean by type identity, says Lewis, is not that perfect duplicates have the same properties *per se*, but that perfect duplicates have the same *intrinsic* properties, whilst they may differ in their *extrinsic* ones.[18]

So far so good, but Lewis admits there is a problem, for the concepts of intrinsic and extrinsic are themselves philosophically troubling. The problem is that there seems to be no way of distinguishing intrinsic properties from their extrinsic cousins without appealing to the concept of a duplication.[19] What we are left with is a tight little circle of interdefinables.[20]

So instead, Lewis approaches the problem from another angle. Rather than beginning with intrinsic properties, he starts with *natural* properties. Natural properties, he states, are properties which are *ex officio* intrinsic.[21] But the term "natural" is not synonymous with "intrinsic", for the converse relation does not hold; not all intrinsic properties are natural (an intrinsic property may be disjunctive, for instance).[22]

Natural properties, he argues, give us a window into the tight circle; for once we have natural properties we can define duplication in terms of the sharing of natural properties, and then define intrinsic properties in terms of those that never differ between duplicates.[23] But not only does it allow us to break into the circle, it also provides a method of analysing qualitative identity. Sameness of type, Lewis argues, is related to the concept of duplication, which is defined in terms of the sharing of intrinsic properties. In linking the concept of intrinsic nature with natural properties, he can then provide a link back to the concept of perfect duplicates, and hence, qualitative identity.

So Lewis has now provided an analysis of qualitative identity, but what are these mysterious "natural" properties? Haven't we just exchanged one metaphysical mystery for another? Lewis believes not. Natural properties, he argues, should be taken as primitive, and thus, are not to be conceptually analysed in terms of anything more fundamental. Lewis defends this move by stating that all ontologies have their primitives, for instance, in Armstrong's case it's the universals which are taken as primitive. If Armstrong is permitted to have primitive universals, why shouldn't natural properties be allowable as primitives?[24]

Given that Lewis has provided a conceptual analysis of qualitative identity, he now needs to provide an identity criterion. For unless such a criterion is supplied, we have no way of *a posteriori* distinguishing natural properties from the multitude of unnatural gerrymanders and disjunctional intrinsics that pervade the Lewisian landscape.

Lewis cannot, however, supply a criterion that involves an appeal to a property's causal or nomological role, for according to the Distinctness Thesis, these roles are only contingently related to a property's identity. But alas, in a fashion that somewhat parallels Armstrong's, Lewis abandons Humeanism and employs a property's causal and nomological role to provide the basis of his analysis of natural properties, and hence, qualitative identity:

> Physics is relevant because it aspires to give an inventory of natural properties — not a complete inventory, perhaps, but a complete enough inventory to account for duplication among actual things. If physics succeeds in this, then duplication within our world amounts to *sameness of physical description*.[25]

Thus, once again it is at this point that the Humean metaphysics becomes unstuck. Lewis's analysis of qualitative identity involves equating sameness of natural property with sameness of physical description. Since physical descriptions are couched in terms of a property's causal and nomological role, Lewis is committed to analysing duplication, and hence, qualitative identity in terms of a property's causal and nomological role. Consequently, as with Armstrong, Lewis's ontology is fundamentally anti-Humean in that the (metaphysically significant) properties in his ontology possess their causal powers *necessarily*.

At this point Lewis might object to the accusation that his analysis entails that his natural properties have their causal powers necessarily, for one crucial difference between his and Armstrong's analysis is that Lewis's analysis of qualitative identity is *world-relative*; thus strictly speaking, all his analysis states is that *in our world* natural properties that possess the same physical description are duplicates.[26]

No doubt this restrictive condition was added to allow for the (supposed) contingency of the laws of nature. For it follows that if the laws of nature are contingent, then the same kinds of natural properties can have different causal powers, and hence, different physical descriptions in different possible worlds. Thus construed, it seems that this extra condition entails that Lewis's world-relative criterion[27] of qualitative identity avoids a commitment to anti-Humeanism.

However, although Lewis can restrict his analysis so it is world-bound in order to accommodate the contingent status of laws, some explanation is needed to justify why, *in our world*, the natural properties that have the same physical descriptions are deemed to be qualitatively identical. For if a natural property possesses its causal powers contingently,[28] why should it not be that, within a world, the same kinds of natural properties all possess different causal powers, or alternatively, different kinds of natural properties all possess the same kinds of causal powers?

One obvious answer is to disallow this state of affairs by appealing to the regularity (or necessity) of laws. Unfortunately however, such an appeal is not sufficient to rule out this possibility, for the simple reason that regular nature of laws is established on *a posteriori* grounds. Consequently, all the regularity of laws guarantees is a consistent correlation between entities possessing certain causal powers (i.e., dispositional properties) and the kinds of behaviour they tend to display. Thus, strictly speaking, the observed necessary connections *cannot* be between natural properties and the behaviour they exhibit, for as we have no *a posteriori* access to the true nature of natural properties (for stripped of their causal powers they are inscrutable), we cannot therefore observe such connections.

For instance, consider two worlds, the first possessing three kinds of natural properties each paired with a different kind of disposition $[(P_1, D_1), (P_2, D_2), (P_3, D_3)]w_1$ and the second world possessing three instances of one kind of natural property each married to a different kind of dispositional property — the same three kinds of dispositional properties as in the first world $[(P_5', D_1), (P_5'', D_2), (P_5''', D_3)]w_2$. From an empirical point of view these two worlds are indistinguishable. This perceived similarity would be reflected in our scientific descriptions; i.e., we would have no reason *not* to believe that they both contain the same kinds of properties and the same kinds of laws. Yet, according to Humeanism, each world is composed of entirely different kinds of entities.

But just as the Humean has no *a posteriori* way of telling these two worlds apart, the Humean has no *a posteriori* way of knowing whether two entities which possess the same dispositions *within a world* are both grounded in the same kind of natural property. For example, consider two entities x and y. Both inhabit the same world, and both exhibit the same set of dispositions. From an empirical point of view, they would appear identical in kind. But given that

Humeans claim that dispositions are only contingently related to the properties that possess them, it is logically possible that the entities *x* and *y* are *different* kinds of natural properties. Moreover, this situation occurring *within a world* cannot be disallowed on the strength of an appeal to the regularity (or necessity) of laws, for the observed regularities are compatible with this state of affairs.

A Humean could circumvent this problem by stipulating that, *of necessity*, within any given world, the same type of natural property must always connect up with the same type of disposition (although in any particular world, which types link up with each other is contingent). But making such a stipulation injects a form of necessity into the analysis that is not permissible. One cannot simply stipulate that a contingent relation holds *of necessity* without explaining why. This is, of course, the problem of the necessity of laws viewed from another angle, and hence, the same problems that beset contingency theories emerge with respect to dispositions. Thus, parallel "solutions" to the problem can be given: either the regular nature of the connection between natural properties and their causal powers is stipulated as necessary (yet mysterious), or else it is taken as a brute fact. However, if it is taken as brute fact as the regularity theorists have it, given that knowledge of such regular connections cannot be gained empirically, on what authority can they make this claim?

So although Lewis's world-relative restriction exonerates his natural properties from being identified with physical descriptions in all possible worlds, it does not absolve Lewis's analysis from the charge of violating the Distinctness Thesis. For if Lewis wants to match natural properties with causal powers within a world, given that such connections are empirically unobservable, and hence there are no "brute regularities" to appeal to, he must stipulate that such an identity holds *of necessity*. But to do this, he needs to inject his metaphysics with some mysterious kind of world-bound necessity to guarantee that this connection holds. Lewis cannot stipulate that this identity holds by way of mere "brute fact", as such a move is philosophically indefensible. For given that such relations are contingent, *stipulating* that they hold *without exception* within a world is tantamount to employing metaphysical magic.

Consequently, even a world-bound analysis of natural properties entails a commitment to some kind of necessary connection between natural properties and their causal powers. Therefore, by identifying natural properties with the inventory of physics within this world, Lewis thereby violates the Distinctness Thesis which by his own admission, forms the cornerstone of his metaphysics. Nevertheless, it is not the violation itself which is a sin, but the repercussions it has for his research programme. The bottom line is that Lewis's Humean metaphysics is fundamental inconsistent;[29] he *cannot* subscribe to the Distinctness Thesis (and the contingency theory of laws it is said to entail) if he embraces his analysis of natural properties so described.

Lewis's commitment to the Distinctness Thesis leaves him in a bind. He needs to provide an identity criterion for natural properties, and on one level, like Armstrong, he (implicitly) acknowledges that the only criterion capable of performing the metaphysical work is a causal one. Yet, identifying the nature of properties with their causal roles is an anathema to a Humean. But without a method for *a posteriori* identifying natural properties Lewis's metaphysics has no empirical credibility.

Lewis has two options: he can acknowledge that the inconsistency in his metaphysics lies in the fact that his natural properties have been engaging in an illicit liaison with a distinctly *un*-Humean causal criterion. Furthermore, he can acknowledge that this metaphysical merger has not been entirely unfruitful, for the illegitimate offspring of this union is an ontology friendly to empiricism. This is a crucial point, as the major shortcoming of a purely Humean ontology is the fact that it is metaphysically barren, for without a causal criterion, it has no empirical relevance. Not surprisingly, Lewis does not want such a metaphysics; however, the central problem for him is that he cannot, *on pain of contradiction*, formally recognise this empirical progeny without first acknowledging the existence of this *un*-Humean liaison. But to do this he must first annul his pledge to the Distinctness Thesis and openly embrace an anti-Humean ontology.

Nonetheless, if Lewis wants to avoid a metaphysical divorce from Humeanism and retain a consistent metaphysics, he does have one other option. In order to keep his natural properties (which do most of the metaphysical work in his ontology) *as well as* the Distinctness Thesis (and the contingency theory of laws it entails) he must abandon the identification of natural properties with the physicists' inventory, and/or supply an alternative criterion for physicality that does *not* involve reference to a property's behaviour. For once he severs the link between *physical* properties and *natural* properties, the inconsistency in his theory dissolves. Strictly speaking, such an identification is not permissible in the first place, for given Lewis's conceptual analysis of natural properties *together* with his commitment to a contingency theory of laws, we would find that when we turn to physics for an inventory of natural properties, there is nothing in the physicists' inventory that can be said to deserve the name.

Is this severing of the link between natural properties and physical properties all bad? Lewis thinks not, stating that it is the natural properties themselves which matter, and not the theory (in this case the relevance of physics) which tells us what they are. For if physics were an utter failure, there would still be duplication in virtue of shared natural properties.[30]

However, I strongly disagree with the spirit of this response. I accept that postulating as primitive, mysterious metaphysical entities such as natural properties may be justifiable on the grounds that they are needed to provide the requisite apparatus to delineate the sparse array of elite properties that do the metaphysical work. However, a metaphysical theory which can theoretically define qualitative identity, yet can offer no *a posteriori* way of utilising this definition, whilst conceptually interesting, is functionally useless. Lewis's natural properties may be primed for metaphysical work, but *ex hypothesi* they are unemployable in an empiricists' metaphysics. All his properties must live on social security. But metaphysical credit notes hard to secure, and despite the proliferation of possibilila, his ontology has no metaphysical currency to pay its way.

Thus a consistent Lewisian ontology is an ontology of properties that, by its very nature, cannot be known *a posteriori*. But if this is the only consistent alternative to an anti-Humean ontology, what reason do we have for embracing it? Such an ontology cannot pay its way by appealing to its explanatory value, for it has none (or very little). Moreover, as an empiricist metaphysics that purports, amongst other things, to provide an analysis of lawhood, it is found to be equally wanting. For the cost of maintaining the Distinctness Thesis, and hence a contingency theory of laws, is a permanent schism between a philosophical and scientific ontology. For the entities that feature in laws, and thus form the basis of a scientific ontology, have no discernible relation to the properties that have nomological significance in a (consistent) Lewisian metaphysics. But if such an unholy divorce between science and philosophy is the price of retaining the Distinctness Thesis, I suspect that very few would be willing to pay it.

Thus for Lewis, the choice is between an inconsistent metaphysics which embraces an un-Humean causal criterion, or else an empirically barren metaphysics which is faithful to the Distinctness Thesis. So in the end, it seems that Lewis's Humean analysis in nominalistic dress fares little better than Armstrong's realist analysis. For apart from their respective ontologies, the only salient difference between the two is that a sphere of inscrutable properties is not an option for Armstrong's *a posteriori* ontology. But this difference aside, in their present formulations, both Lewis's and Armstrong's theories are fundamentally inconsistent in that they both violate the central principle on which their metaphysical theories are based.

4. HUMEAN SUPERVENIENCE AND ONTOLOGICAL REDUCTION

The discussion so far has focussed on an ontology composed of properties and relations. However, in our world these properties and relations do not seem to constitute a single flat layer. Whilst properties may be simple, they can also be complex in that they are composed of an aggregation of more simple properties. These complex properties taken as individuals may then form new relations with other individuals. In effect, what we have is a hierarchy of properties and relations which increases in complexity the further we ascend the metaphysical structure.

Thus construed, properties and relations constitute the metaphysical bricks and mortar. But a metaphysics which postulates primitive properties and relations and leaves it at that has unfinished business. What we also need is a recipe for building the metaphysical edifice. What we need is an analysis of how the different levels of the hierarchy are connected to each other.

The concept which is employed to do this metaphysical work is that of "supervenience". This concept designates some kind of relational property, but the kind of relation is highly specific; it is a non-causal relation of *dependence*, and herein lies the significance of the concept in philosophy. A supervenient relation performs an important unifying role in philosophy for it constitutes a bridge, both metaphysically and explanatory, between seemingly diverse and disparate properties. For example, when aestheticians say that aesthetic merit supervenes on the arrangement of the pigments on the canvass, they are offering some kind of analysis of the relation of dependence between aesthetic merit and the physical attributes of the work of art. In effect, they are stating that the aesthetic merit is dependent on, and is in some way determined by the underlying (i.e., subvenient properties). Similarly, when a physicalist states that the mind supervenes on the brain, the claim being made is that there exists a tight non-causal relation between the mind and the brain in that the mind is in some way dependent on the brain for its existence.

In recent years much philosophical ink has been devoted to formally explicating the concept of supervenience.[31] The result of these philosophical labours has been a proliferation of definitions each claiming metaphysical supremacy — and a lot of confusion. In addition, an analysis of the supervenience relation is further complicated by the fact that it is also related to two other concepts: *viz*, emergentism and reductionism.[32]

Generally speaking, reductive supervenience has popular appeal and no doubt this stems from the fact that concept of reductionism is seen to be relatively unproblematic. The same, however, cannot be said for emergentism. It has been claimed that emergentism's great advantage is that it has the most potential for providing insightful explanations into the ontological origins of a diverse array of phenomena. But for many philosophers, genuine emergence seems too much like an ontological free lunch. Emergent properties *by definition* do not reduce to the sum of the subvenient component parts. So how do the emergent properties come about?

In essence, the problem with emergent properties is that they are seen to be *ex hypothesi* unexplainable. But, it is argued, for a supervenience relation to possess metaphysical value, it must be capable of providing an explanation to account for the connection between the supervenient property and the subvenient base. Emergent supervenience, by hypothesis, fails this criterion.

But reductive supervenience, it is argued, *can* offer an explanation. And the cherished tool of those defending ontological reduction is the process of *intertheoretic* reduction. Intertheoretic reduction, it is argued, functions as an explanatory mechanism for fleshing out how the ontological reduction of one physical property to another is possible.

Basically intertheoretic reduction occurs when a theory T_1 can be reduced (with the aid of bridge laws) to another theory T_2. A paradigm example of this is the (as yet unrealised) intertheoretic reduction of chemistry to physics. For generally speaking, chemistry and physics employ a different ontological vocabulary. But if we are to link these two vocabularies, we need a translation manual. Intertheoretic reduction, it is argued, supplies this manual in the form of bridge laws that have the form of nomological bi-conditionals which connect a term (or set of terms) from the vocabulary of one theory to another. Thus by using intertheoretic reduction, an explanation can be offered that illustrates how one level of entities on the ontological hierarchy is related to another set on another level.

However, for intertheoretic reduction to have any relevance to ontological reduction, we need the *additional* assumption that the hierarchy of *theories* that can be intertheoretically reduced, maps on to the hierarchy of properties being *ontologically* reduced. It is here that the Humean ontology again becomes unstuck. In reality *intertheoretic* reduction is just *nomological* reduction parading under another name. It is the reduction of one set of entities with a defined set of nomological roles, to another set of entities with a different set of nomological roles. But this situation implies that Humeans are not allowed to make a strict connection between the ontological hierarchy and the nomological hierarchy. Or at the very least, any mapping made between the two must be contingent.

Furthermore, as the Humean divorces the nature of properties from the laws they instantiate, utilising intertheoretic reduction to explain ontological reduction provides the Humean with a dilemma. If the link between the ontological and nomological structure is contingent, does this then imply that ontological reduction is contingent also? Is it the case that a property can supervene on a set of subvenient properties in one world and not another? Or should we say that ontological reduction occurs across all worlds, and it is only the *behaviour* that these supervenient and subvenient properties instantiate that can differ from world to world?

Take, for example, the following situation. Given that *in our world* protons are composed of three quarks (red, green and blue), should we say that the natural property or universal "proton" supervenes on these quarks *necessarily* (i.e., the supervenient relation holds in all possible worlds), or is the supervenience *contingent* (i.e., it holds in some worlds but not others)?

The Humean can choose one of the following options:

(a) *Necessary Ontological Reduction*. Protons supervene on quarks necessarily, (proton = quark + quark + quark *in all worlds*). However, due to the contingent nature of laws, protons and quarks in worlds with different laws from ours will behave differently.

Or:

(b) *Contingent Ontological Reduction.* Due to the contingent nature of laws, protons supervene on quarks in some worlds and not others.

Which response should the Humean embrace? The reply that is favoured ultimately depends on the criterion adopted with respect to the problem of type identity. If you do not wish qualitative identity to be intrinsically linked with causal powers and the role a physical property plays in a nomological system, the first option — *necessary* ontological reduction — is the plausible alternative to embrace.

However, it is by no means a credible alternative, for as was argued earlier, divorcing a property's physical identity from its causal role results in an ontology of physical properties which are *ex hypothesi* unknowable, and therefore indistinguishable from each other. The problem is *not only* that we are unable to recognise a proton *as a proton* in another world with different laws (hence those embarking on metaphysical interworld excursions should not expect any identifying landmarks), but this analysis also implies that we have no way of recognising protons as protons *in this world.* In fact, we have no way of distinguishing protons from electrons, quarks or indeed, any other physical phenomena. So although ontological reductive supervenience across all worlds is internally consistent, its practical shortcomings are so gross that there is little to recommend it.

However, the second alternative, contingent reductive supervenience, also lacks any redeeming features. In fact, it is logically inconsistent with the Distinctness Thesis, thus is essentially anti-Humean in spirit. It violates the Distinctness Thesis, for in order to make sense of the spirit of the reply, we need to identify the proton with the kind of causal properties it manifests. In effect, what we are being asked to consider is a proton *as we know it,* that cannot be decomposed into a structure consisting of three quarks.

But to consider a proton in this way, we need to have some knowledge of its properties. However the Distinctness Thesis appears to prohibit *a posteriori* knowledge of protons, and yet the Humeans have not supplied any method of making an *a priori* identification of it. Thus it appears that if we are to make sense of contingent reduction, we *must* then identify physical properties with their causal powers. But we can't, for this violates the Distinctness Thesis, and thus entails abandoning the contingency theory of laws. Once we do this we are logically committed to the claim that protons supervene on quarks in all possible worlds (where protons exist, that is). But this contradicts the claim that the supervenience is contingent. The moral: contingent reductive supervenience in Humean dress is fundamentally contradictory.[33]

With respect to Humean supervenience doctrines the choice, it seems, is between inconsistency and inscrutability. Choose *necessary* ontological reduction, and you can retain a consistent metaphysics, but you must marry it with a esoteric ontology. If you prefer to have an ontology that permits itself to be known on more intimate terms, choose *contingent* ontological reduction. But be warned, in this instance, intimacy comes at a price. The cost is infidelity, for such an ontology must, on pain of contradiction, betray the principle on which contingent supervenience is founded, *viz.* the Distinctness Thesis.

In theory, neither of these options are commendable, but from a practical point of view, a Humean must subscribe to one or the other of them. However, the situation where a philosopher consistently embraces one or the other is rarely found. Instead, what usually occurs is a form of metaphysical 'double-think', where both options are individually subscribed to in different situations.[34]

Although both Humean supervenience theses are philosophically unacceptable, they do have their individual merits. Necessary reductive supervenience is friendly towards the Distinctness Thesis, and thus is usually embraced when a Humean wishes to emphasise this principle. Contingent reductive supervenience in a Humean framework, although inherently contradictory, *sounds* friendly to both Humeanism and empiricism. Thus when Humeans want to link their ontology with the world as we know it, they tend to embrace this second option. The fact that both options employ a different criterion of type identity, and therefore are mutually exclusive, does not seemed to have been noticed.

As can be seen, the problems associated with Humean supervenience exactly mirror the problems inherent in a Humean ontology. This explains why so much of the literature which discusses supervenience from a Humean perspective is fundamentally incoherent. The proliferation of supervenience definitions that come complete with a catalogue of occasions stating when each type applies, is symptomatic of an attempt to make sense of this incoherence. However, *more* definitions and a more refined catalogue will not purge the theory of its contradictions. The problem lies in the fact that the concept of Humean supervenience is *itself* incoherent, and the reason why Humeans cannot coherently analyse reductive supervenience is directly linked to the fact that the ontology that forms the basis of Humeanism is *itself* inherently inconsistent. Hence any ontological edifice constructed according to the principles of Humean supervenience is, in effect, jerrybuilt.

The solution, I urge, is to reject the Distinctness Thesis and the metaphysical baggage that accompanies it. However, I realise that anti-Humeanism might not be an option palatable to everyone. But those Humeans who find that they cannot betray their vows to the Distinctness Thesis must relinquish their allegiance to empiricism, and retreat from the world of a physicists' ontology to reside in a realm of inscrutable matter — a kind of ontological heaven if you like — shrouded in a veil of causal inefficacy, where no property need perform any metaphysical work.

However, those who choose not to inhabit this Humean castle-in-the-air must ground their metaphysics on a more solid empirical base. But, as I have shown, an empirical terrain cannot consistently support a Humean ontology. Thus I conclude, the House of Hume rests on unsound foundations, and therefore should be condemned as unfit for metaphysical habitation.

La Trobe University, Bundoora, Victoria

NOTES

* I wish to thank David Armstrong and an anonymous referee for helpful comments.

1 What are "distinct" existences? Lewis (1986) pp.88–90, suggests that they are things which occupy distinct spatio-temporal positions. Armstrong (1989) p.20, defers judgement on this issue, and although he states that it may be possible for two distinct existences to occupy the very same position, he nevertheless agrees that Lewis's analysis qualifies as a sufficient condition. Both Armstrong and Lewis subscribe to Hume's Distinctness Thesis.

2 Those offering *anti*-Humean (also known as "essentialist") ontologies include Bigelow *et. al.* (1992), Ellis and Lierse (1994), Fales (1986) & (1990), Harré (1970), Harré and Madden (1973) & (1975), Shoemaker (1980), and Swoyer (1982). However, it should be noted that the arguments advanced in this paper are against Humeanism *per se,* and thus this paper is neutral insofar as it does not presuppose a commitment to a particular species of *anti*-Humeanism.

3 Apart from being employed to defend a contingency of laws, the Distinctness Thesis is also used as a principle of recombination. For example, see Armstrong (1989) p.20 and Lewis (1986) p.87.

4 The shortcomings of a Humean theory of laws are well documented, and for this reason I will not rehearse them here (e.g., see Armstrong (1983)). Most adherents to the Distinctness Thesis seek to smooth over these deficiencies by finding some other way of injecting laws with the requisite necessity. Such amendments have met with varying degrees of success, but always at a cost. I suggest that the problems that beset contingency theories of laws are symptomatic of the inadequate Humean ontology that forms the basis of such theories, thus no amount of tinkering with the analysis of lawhood alone will fix them.

5 I will not attempt to offer an analysis of causality in this paper.

6 This is a necessary, but not sufficient condition. See section 3 for a more detailed criteria for physicality.

7 Strictly interpreted, this criterion implies that a property is never physical *simpliciter*. For the Distinctness Thesis is compatible with the situation where a property is causally efficacious in one world, and yet causally idle in another. This then implies that the characteristic of physicality is *world-relative*. So photons (say) qualify as physical properties in some worlds, and not in others. Whilst this in itself is not an objection, I am not sure whether all Humeans would feel comfortable with this.

8 Armstrong reserves this appellation for those who adopt a regularity theory of laws.

9 Armstrong (1978) vol 2, p. 11.

10 Armstrong (1978) vol 1, p.133. Although in his (1978) vol. 2, ch. 16 where he discusses the identity conditions for universals in more detail, Armstrong asserts that the justification for this identity criterion is merely pragmatic. However, I do not believe that the formal/pragmatic distinction is particularly relevant to the point being made. What matters is the *kind* of identity condition being postulated, and not whether the justification is formal or pragmatic.

11 See Armstrong (1983) pp.138–39.

12 A propensity need not be analysed purely in terms of a subjunctive conditional, as Armstrong would have it, but it can be interpreted as a genuine *occurrent* property in virtue of which the subjunctive holds (see Ellis & Lierse (1994)). This avoids the absurd situation of a physical entity being characterised as nothing more than a cluster of subjunctive conditionals.

13 For further arguments supporting this point of view, refer to the *anti*-Humean literature cited above.

14 In particular, see Lewis (1983a) & (1986).

[15] Lewis (1983a) p.356 & (1986) p.60.
[16] See Schlesinger (1963) for a discussion of these two general principles.
[17] Lewis (1983a) pp.355–57 & (1986) pp.59–62.
[18] Lewis offers the following analysis: "Two things are duplicates iff they have precisely the same intrinsic properties, however much their extrinsic properties might differ." ((1983a) p.355.)
[19] Lewis's analysis of the intrinsic/extrinsic distinction is as follows: "Property *P* is intrinsic iff, for any two duplicate things, not necessarily from the same world, either both have *P* or neither does. *P* is extrinsic iff there is some such pair of duplicates of which one has *P* and the other lacks *P*." (Lewis (1983a) p.355.)
[20] Lewis (1983b).
[21] Lewis (1983a) p.357.
[22] Lewis (1986) p.61.
[23] ibid. p.89.
[24] ibid. pp.63–66.
[25] Lewis (1983a) p.357. Italics mine.
[26] Nevertheless, there is a problem with Lewis's analysis of duplication insofar it only applies *within* a particular world. For Lewis claims that his natural properties are *intrinsic* properties; thus they are the same in all worlds. But if the only way we can know them and refer to them is by their dispositional natures (which differ from world to world), then Lewis is stuck with the situation of having to identify such properties in terms of their contingent properties. As an identity criterion, this is a little strange. One consequence of this is that we have no way of referring to other-worldly natural properties except by indexing the descriptions of such properties to this world. Furthermore, we would have no way of *recognising* the same natural properties in different possible world where they are governed by different laws.
[27] However, given that this analysis of qualitative identity only applies *within* a world, Lewis still needs to supply a transworld analysis of qualitative identity. For such an analysis is a necessary prerequisite for his analysis of concepts such as supervenience, laws and causation. (See Lewis (1983a) pp.359–70.)
[28] There is the further question as to whether this world-bound analysis is consistent with Lewis's claim that natural properties are the ones that "carve reality at the joints", and are those "whose sharing makes for resemblance, and the ones relevant to causal powers" ((1983a) pp.346–47). No doubt this characterisation of natural properties is meant to delineate natural properties from the abundantly many other irrelevant properties, and to apply in all possible worlds ("A property is natural or unnatural *simpliciter*, not relative to one or another world." Lewis (1986) p.61.)) But if natural properties are to be divorced from their causal/nomological roles to accommodate the contingent nature of laws, then natural properties can only be causally relevant and salient to duplication *within* a world.
[29] As shown, Lewis's world-bound analysis of natural properties also entails that his metaphysics has built-in modal primitives — but why this is a problem is another story.
[30] Lewis (1983a) p.357.
[31] In recent times there have been many papers fully (or partially) devoted to this topic. An interesting collection can be found in the *Southern Journal of Philosophy*, supplement to volume 22, 1984, (The Spindel Conference on Supervenience), and in Beckermann, *et. al.* (eds.) (1992). See also Lewis (1983a).
[32] For an interesting discussion of these two concepts see Beckermann, *et. al.* (eds.) (1992).
[33] This is not to say that in an anti-Humean ontology it is not possible for a supervenient property to have more than one possible subvenient base. For example, "being copper" is a supervenient property that has a range of possible subvenient bases (e.g., the different copper isotopes, different ionisation and oxidation states, etc.). And although copper is defined in terms of its causal/nomological role, in such a case, the differences in the subvenient bases are deemed to be non-essential properties of copper. "Being copper" is defined purely in terms of physical properties, namely of being composed of atoms containing 29 protons.

But there is also another way in which a property can have many possible subvenient bases, and that is where the identity of the property is not defined in terms of its physical/nomological properties, but broadly defined in terms of its functional role. For example, tables, chairs, vases, etc. are complex properties that supervene on the nature and arrangement of their subvenient bases, but what determines whether something is a chair, for instance, is not primarily what it is made of, but the role we ascribe to it. The Humean can, of course, use this second type of contingent reduction, but the Humean *cannot* use contingent reduction as an analysis of natural properties, for as they are intrinsic *ex officio,* they are exact duplicates in all worlds.

[34] For example, see Lewis (1986) p.169, where he states that it is possible to *define* a proton in terms of its structural components (i.e., quarks), and compare this with p.159, where he postulates the existence of an impoverished possible world where protons are indivisible particles. (See also Lewis (1986) pp.161–163 & Teller (1984) p.151.) Further examples of such conflicting analyses can be found in Humean discussions of structural universals, where structural properties are *defined* in terms of their structural components (e.g., the structural universal "methane" is defined as consisting of one carbon and four hydrogen universals). But when Humeans turn from the subject of ontology to analysing dispositions and laws, they countenance the possibility of such states of affairs where electrons, for instance, lack the property of negative charge. Clearly there is an inconsistency between these two conceptions.

REFERENCES

Armstrong, D. M., (1978) *Universals and Scientific Realism,* (2 vols.) Cambridge, Cambridge University Press.

— (1983) *What is a Law of Nature?,* Cambridge, Cambridge University Press.

— (1989) *A Combinatorial Theory of Possibility,* Cambridge, Cambridge University Press.

Beckermann, A., *et. al.* (eds.), (1992) *Emergence or Reduction?: Essays on the Prospects of Nonreductive Physicalism,* Berlin, de Gruyter.

Bigelow, J. C., Ellis, B. D. and Lierse, C. (1992) 'The World as one of a Kind: Natural Necessity and Laws of Nature', *British Journal for the Philosophy of Science* 43, 371–388.

Ellis, B., & Lierse, C. (1994) 'Dispositional Essentialism', *Australasian Journal of Philosophy* 72, 27–45.

Fales, E. (1986) 'Essentialism and the Elementary Constituents of Matter', *Midwest Studies in Philosophy* 11, 391–402.

— (1990) *Causation and Universals,* London, Routledge.

Harré, R. (1970) 'Powers', *British Journal for the Philosophy of Science* 21, 81–101.

Harré, R., and Madden, E. H. (1973) 'Natural Powers and Powerful Natures', *Philosophy* 48, 209–230.

— (1975) *Causal Powers: A Theory of Natural Necessity,* Oxford, Blackwell.

Lewis, D. K. (1983a) 'New Work for a Theory of Universals', *Australasian Journal of Philosophy* 61, 343–377.

— (1983b) 'Extrinsic Properties', *Philosophical Studies* 44, 197–200.

— (1986) *On the Plurality of Worlds,* Oxford, Basil Blackwell.

Schlesinger, G. (1963) *Method in the Physical Sciences,* London, Routledge.

Shoemaker, S. (1980) 'Causality and Properties' in P. van Inwagen (ed), *Time and Cause: Essays Presented to Richard Taylor,* Dordrecht, Reidel, pp. 109–135.

Swoyer, C. (1982) 'The Nature of Natural Laws', *Australasian Journal of Philosophy* 60, 203–223.

Teller, P. (1984) 'A Poor Man's Guide to Supervenience and Determination', *Southern Journal of Philosophy,* Supplement to Volume 22, pp. 137–162.

PETER FORREST

PHYSICAL NECESSITY AND THE PASSAGE OF TIME *

The basic idea of this essay is that the past is no longer contingent, hence that which has always been past was never contingent, and is, therefore, necessary in a time-independent fashion.[1] In this way we can give an account of physical necessity in terms of the passage of time. My purpose is to clarify this somewhat obscure idea, and to argue for the resulting account of necessity.

1. VARIETIES OF NECESSITY

The argument which I present requires a number of premisses which could be questioned, but will not be on this occasion:

(1) There is a modality of physical necessity.

(2) It is objective, in the sense that what is physically necessary is quite independent of what human beings are, or do.

(3) It is precise.

(4) It is to be distinguished from the modality of metaphysical necessity.

(5) Merely possible things (including uninstantiated universals) either do not exist or else depend for their existence on actual things (including instantiated universals).

Although detailed argument for the occurrence of a modality of physical necessity is beyond the scope of this essay, it might be helpful to sketch a reason for believing this. It is that it is either ineliminable from physics or else eliminable only in *ad hoc* ways. While chemistry and zoology are primarily concerned with the classification and understanding of actual cases, physics is concerned with what is physically possible. For instance, no exposition of classical mechanics would restrict attention to those position-momentum pairs which actually occur, by avoiding momenta which were too large, say. Nor, on the other hand, would we usually spend much time discussing states which are mere metaphysical possibilities, such as those of particles in a four-dimensional Euclidean space in which the familiar inverse square laws are replaced by inverse cube laws. Instead, every theory in physics is, primarily, a description of the physically possible.[2]

The challenge is to say what sorts of items are physically necessary and to explain their necessity. The semi-Humean proposal would be that various regularities are physically necessary. There are some *prima facie* difficulties with this proposal. One is that these regularities only hold in hypothetical idealised or simplified situations.[3] Related to this is the point that for the most part the regu-

P.J. Riggs (ed.), Natural Kinds, Laws of Nature and Scientific Methodology, 49–62.

larities admit of exceptions. Therefore they must be described using a 'for the most part' or 'other things being equal' qualification, with the result that any way of making them precise would be physically contingent. For example, the regularity that massive bodies attract each other (with such and such a force) only holds in the absence of strong repulsive forces such as those between like charges. Again, the regularity that fifty per cent of free neutrons will decay in 898 seconds, will hold of a given collection of, say, 100 neutrons, only by chance and so is physically contingent.

As far as I can see, the way round this difficulty is to posit various entities such as gravitational forces, in the first example, or propensities, in the second. Then there may indeed be exceptionless regularities. In this way we obtain the first proposal for what is physically necessary, namely:

(i) Regularities concerning forces or powers or propensities,[4] for example, the regularity that all massive objects have the power to attract other massive objects, or that every neutron has a certain propensity to decay.

Since, however, we seem forced to posit some theoretical entities, such as forces, powers or propensities, in order to give an account of what is physically necessary, there is no reason to reject more explicitly metaphysical proposals about what is physically necessary, such as the following:

(ii) A certain relation (of necessitation or probabilification) between universals.[5]

(iii) The kind of universe we live in.[6] In this case, various regularities hold, in whatever idealised fashion they do, because this is the sort of universe which is, among other things, characterised by just such regularities holding.

(iv) Accessibility relations between possible worlds.[7] On this account, what is physically necessary is what holds in all worlds accessible to ours. Here accessibility supervenes on the characteristics of the worlds in question.

(v) The underlying symmetries of the situation.[8] This might find some support in the attempts at a unified field theory, which are based on the search for symmetry.

In all cases we can give the following account of what it is for a generalisation to *manifest* the physically necessary item being posited:

Relative to background information and detailed knowledge of the item in question, and nothing else, the epistemic probability of any instance of the generalisation is appropriate.

Where the generalisation is universal, an appropriate epistemic probability is a high one. For statistical generalisations an appropriate epistemic probability is one which approximates the proportion indicated in the generalisation (about 50% in the example of the half life of the neutron).

The problem of explaining the physical necessity of whatever is manifested by the generalisations discovered by science is compounded by the widespread intuition that in addition to physical necessity we have a modality of metaphysical necessity. To complicate matters further, there seem to be two ways of characterising metaphysical necessity, the non-actualist and the theistic.

A realist about non-actualities, whether thought of as particulars (parts of possible worlds, as in Lewis' account) or universals[9] may characterise the metaphysically necessary as that which is true of all non-actualities as well as the actual.[10] Given this characterisation there are many physically impossible but metaphysically possible items which would be manifested by generalisations about familiar properties. Indeed something which would manifest itself in the statement of Newton's law of gravitation would, no doubt, be an example, since that law is false.

There is an alternative, theistic, characterisation of metaphysical necessity. Any non-creature, that is, anything which exists independently of creation, is metaphysically necessary in the theistic sense. It is widely assumed that the kind of universes that exist and the laws which hold in them follow from rather than constrain the act of creation. So theistic metaphysical necessity is stronger than physical necessity.

Our task, then, is to seek the best theoretical account of physical necessity, in such a way as will distinguish it from metaphysical necessity. The account I give will be a version of theistic metaphysical necessity, although atheists could adopt an *as if* variant, according to which there is a 'time before time' prior to the *as if* creation of the universe as we know it. One reason for preferring either the theistic characterisation of metaphysical necessity or its *as if* variant over the non-actualist account is that they are more obviously consistent with my Premiss (5), which expresses a moderate actualism.

2. EXPLAINING PHYSICAL NECESSITY

Of the five suggestions as to what is manifested by the generalisations discovered by scientists, the last, namely the underlying symmetries account, makes no attempt to explain physical non-contingency. That is scarcely surprising since its advocate, van Fraassen, explicitly denies that there is such a thing as physical necessity. In this essay I am, however, assuming without argument that we do need that modality. On the account I shall provide there is in fact no difficulty in taking the symmetries as physically necessary, if we are so inclined.

The first four suggestions all might seem to go further and explain why there is physical necessity. Thus the thesis that the physically necessary items are various regularities concerning forces or powers/propensities may be developed so that exerting these forces, or having the powers/propensities is essential for the kind of entity being considered.[11] Thus to be an electron a particle must have electronish powers. Let us call this the *kinds of entity* explanation.

Again, if the generalisations discovered by scientists manifest relations between universals, then it can be argued that they must hold in all those possible worlds in which those universals are instantiated, which gives them a kind of necessity. The reason for this is that a universal is, by definition, the very same in all its instances and the very same items cannot both be related a certain way in one possible world and not related that way in another. Let us call this the *relation between universals* explanation.[12]

Yet again, ours would, it is said, be a different kind of universe if the 'laws' did not hold.[13] So given that we live in this universe it follows of necessity that the 'laws' hold. Let us call this the *kind of universe* explanation. Finally, that something holds in all accessible worlds is the standard semantics for necessity in a modal logic. So we have an accessibility explanation of physical necessity.

In order to prepare a niche for my own proposal I shall now discuss a quite general difficulty for explanations of physical necessity, which I call the *uninstantiated kinds problem*. The problem is that among kinds of entity which have no (actual) instances we can distinguish between the physically impossible (but metaphysically possible) and the physically possible. An example of the physically impossible kinds are the ninety-niners. They are much like ordinary particles except that for one percent of the time they behave like the antiparticle. So a ninety-niner electron will, now and then, behave like a positron. They do not exist. What is more, I am fairly sure they are physically impossible.

Physically possible but uninstantiated kinds occur if there is some physically possible process which would have led to such kinds but did not. For instance had the Big Bang been more powerful there might have been particles more massive than any which actually have occurred. Or perhaps it is the result of some random occurrence during the Big Bang that there are these rather than other kinds of particle.

The problem, then, is explaining why some kinds are physically possible and others not. My insistence that we are considering uninstantiated kinds is intended to exclude the solution that precisely the instantiated kinds are physically possible. Before proceeding I shall deal with four objections to the way I have set up the problem. First, I seem to assume realism about uninstantiated kinds. But that is just for convenience of exposition. The problem is that it is physically possible there be instantiated kinds, which kinds either actually have no instances or actually do not occur.

The second objection is that I have already solved the problem by the way I described the examples: we can refer back to the history of how the kinds occurred to explain why they are physically possible. I agree, and that is, basically my proposal. If that solution might seem obvious to some then it is surely worth making it explicit, which is what I do in this essay.

The third objection is that I should be considering fundamental kinds, and that my story of the kinds of particle that come into existence shows that they are not fundamental kinds. Perhaps the only fundamental kind is some unified field. My response to that objection is that there might well be no fundamental kinds, but each kind is in some fashion constituted by other, more fundamental ones.[14] In that case we cannot restrict our attention to the fundamental kinds. This reply, however, reinforces the fourth objection, namely that I have no right to assume that my examples are even physically possible, let alone actual: they are merely epistemically possible. What I should say is that *for all I know* there might be physically possible uninstantiated kinds, and *for all I know* there might be no fundamental kinds. I grant that there is something to this objection, but I submit it is at least a desideratum for an explanation of physical necessity that it not depend on the sort of physical theory which is in fact correct, but rather holds for any which, for all we know, might be correct.

Let us now examine how well the various proposed explanations of physical necessity handle the uninstantiated kinds problem. One solution available on all the accounts given is a selective realism about uninstantiated kinds. According to this solution there really are uninstantiated kinds, but not everything we can conceive of as a kind really exists, only some do. And the kinds which do exist, which might or might not have instances, may be identified with the physically possible kinds. This solution, has, however, two defects when compared with my proposed explanation. The first is that because I am distinguishing metaphysical from physical necessity we would then need some other account of metaphysical necessity. My proposal is for a unified account of both sorts of necessity. The second defect is that I am supposing a moderate actualism as expressed by Premiss (5). We are owed, therefore, an account of how the uninstantiated kinds depend on instantiated ones.

We might well, therefore, seek alternative solutions to the uninstantiated kinds problem. On the relation between universals explanation there will be various exclusion laws[15] which might be invoked to exclude various kinds as physically impossible. Those not thus excluded would be physically possible. Thus perhaps some law might tell us that ordinary electrons exclude the ninety-niners. This is, however, a rather uneconomical proposal. For we would need many, perhaps infinitely many, exclusion laws to exclude all the physically impossible kinds. Moreover, it is hard to account for the exclusion laws except by treating them as relations between the excluded kind and the one which does the excluding. This, therefore, requires realism about various physically impossible kinds, such as the ninety-niners, which is even less intuitive than realism about various uninstantiated but physically possible kinds.

The accessibility explanation may conveniently be discussed at the same time as the kind of universe explanation. For, I ask, upon what features of universes does the accessibility relation supervene? If we do not take into consideration what kinds of universe belong to the possible worlds, then there will be no sharp division between the accessible worlds and inaccessible ones and hence physical necessity will not be a precise modality. In addition we shall allow as physically possible those situations which are superficially similar to the actual, say that in which in 1 in 10^{50} occasions a ninety-niner is generated instead of an electron. Surely such a minor deviation can only be excluded as physically impossible by insisting that a world in which that occurred would, in spite of its similarity in all other respects, differs *in kind* from ours. It follows that accessibility must be taken to supervene on the kinds of universe making up the worlds being considered. It is for that reason that I consider the accessibility explanation as a variant of the kind of universe explanation, to which I now turn.

On the kind of universe explanation we might say that the physically possible kinds are just those which occur in universes of the same kind as ours. As an explanation this seems back to front. Surely the kinds of universe are, in part, specified by the kinds of entity occurring in them. It is not the other way round. Thus a universe is of our kind partly because it has our kind of particles in it, electrons, protons and so on. It would seem, therefore, that we need to explain why various kinds of entity are physically possible in order to explain what features are essential to being a universe of our kind.

3. NECESSITY AND THE PASSAGE OF TIME

We cannot remove all mystery, but we make progress if we can show that what seemed to be several mysteries are just the one mystery. Given the difficulties with various attempted explanations of physical necessity we have reason to favour an account which explains both the passage of time and physical necessity using a single unexplained hypothesis. (In addition, we shall be able to explain metaphysical necessity.) This is the thesis of the plurality of actual worlds, by which I do *not* mean that there are many universes unrelated to each other spatiotemporally. Indeed, for the sake of exposition, I shall suppose there is only one universe, which contains everything physical. If there are in fact many separated universes then what I say of ours will hold for each universe. I am suggesting that, even for this one universe, there are many ways it actually is. I call these ways our universe is the *actual worlds*, just as ways the universe might be would be called possible worlds. I use the terminology of worlds in order to borrow the familiar 'apparatus' of possible worlds. It should be remembered, however, that all I mean by a world is a (possible or actual) state of our universe.

I shall be considering locations in spacetime, but, for convenience, I shall talk as if I was considering time in abstraction from space. Hence, when I consider something earlier than *t*, I should be interpreted as meaning whatever is in the backward null cone from *t*.

For each moment *t* in time, there is, I hypothesise, the world which is actual at that moment *t*. What is actual at *t* does not, however, concern only what is happening at or near *t*. For, it is commonly assumed, all that is earlier than *t* is actual at *t*. Moreover if *t'* is later than *t*, then anything which, given what occurs at *t*, must, of physical necessity, occur at *t'* is already actual at *t*. To modify Aristotle's example, suppose that of physical necessity, given that things are as they now are, there will be two fleets at war off the coast tomorrow. Then it is now actual that the fleets will be there, but, we may further suppose, it is not yet actual that there will be a sea battle, nor yet actual that there will not. It is also now actual that either it will be actual that there was a sea battle or it will be actual that there was not. In that case we can, contrary to the way Aristotle is usually interpreted, identify what is now true with whatever will actually be the case and say that although it is not now actual that there will be a sea battle (or now actual that there will not be) it is now true that there will be a sea battle (or now true that there will not be). This idea of time-dependent actuality need not, therefore, be identified with time-dependent truth.

Now, the past cannot be affected — at least not by us, using normal means at the macroscopic level.[16] I submit that our concept of the necessary is primarily of that which we cannot control. It is entirely natural, therefore, to think of the past as necessary. But, unless we are determinists, the past was not necessary when it was the future. It seems therefore that time-indexed necessity and time-indexed actuality coincide. At least that is my hypothesis. What is necessary at a given time is precisely what is actual at that time and the physically necessary is just that which is actual at all times.

My proposal, then, is to accept without explanation the plurality of actual worlds and characterise necessity not as that which is true at all possible worlds, but as that which is true at all actual worlds. My justification for believing in the plurality of actual worlds is that it explains the otherwise mysterious passage of time.

The passage of time is the sort of intuitive datum which should only with reluctance be treated as an illusion. As far as I know there are two reasons which might be given for thus treating it. The first is the appeal of a Parmenidean view which spatialises time and hence implies that the passage of time must be illusory. The second is the supposed incoherence of the passage of time. Much could be, and has been said about the merits and shortcomings of the Parmenidean view. It suffices to say that the chief arguments for it, in my opinion, are its coherence with Einstein's theories and the advantage, when it comes to accounts of change, of thinking of persisting entities as made up of temporal parts, and so assimilating time to space.[17] My aim is to give an account of our passage in time, or, more accurately our growth into the future, which is in agreement with as much of the Parmenidean view which is required to retain those advantages. Replacing the passage of time by our growth into the future is, I grant, not a trivial matter. I shall argue, however, that this replacement requires the rejection only of some negative intuitions, which we should never have trusted.

The supposed incoherence of our passage in time is based on the claim that for us to pass through time there must be a hypertime, that is a higher-dimensional time. And then, it is asked, what explains our passage through hypertime?[18] Even if this is coherent, surely it is uneconomical to hypothesise a hypertime. In response to these objections, I submit that we should distinguish between the temporal dimension and the temporal ordering. The temporal dimension is just that specified by the parameter *t* in physical theories. It is this which we should indeed spatialise, thinking of both inanimate objects and we ourselves as extended in time much as in space, thus obtaining all the advantages of the Parmenidean view. But in addition to the temporal dimension there is, I submit, a temporal ordering, which, in the next section, I identify with the ordering of the actual worlds as more or less determinate.

First, however, let us now consider a traveller's passage through space, say a passage to India. The passenger has different spatial locations at different points in the temporal dimension. But in addition to her 'motion' in that Parmenidean sense there is a change with respect to the temporal ordering. That ordering I treat as the change in what is actual. In accordance with my insistence on the actuality of the (macroscopic) past, I adopt the snowball theory that the actual changes by accumulation rather than alteration[19]. Thus more will be actual tomorrow than is today. In particular, assuming the captain of the boat decides the path from day to day to avoid squalls, the path grows as more of it becomes actual.

As a consequence there is not merely the variation of spatial location with location in the temporal dimension, there is also a growth of the actual path as more becomes actual. I submit that we pass through time in the sense that we inhabit more of the temporal dimension at later 'points' in the temporal ordering. There is an illusion indeed, namely the illusion of the loss of the past. Our memories may fade but the past is never lost from actuality. Hence it is more accurate to say we grow into time than pass through it. The intuition that we pass through time, rather than grow into it, is entirely due to the illusion of the loss of the past. As such it is a negative intuition: the absence of an awareness is mistakenly treated as *the awareness of an absence*. The unreliability of such negative intuitions has been argued by Armstrong using his example of the 'headless woman'.[20]

4. THE ORDERED SET OF ACTUAL WORLDS

I introduced the basic idea of a plurality of actual worlds by considering what was actual as time-dependent. But I now proceed the other way round. Whereas David Lewis proposes a plurality of non-overlapping worlds, I propose a plurality of worlds which can *overlap* in that the same objects can exist in two different worlds. These worlds form the system of actual worlds which is partially ordered in that some worlds are more determinate than others. The temporal ordering is to be identified with the ordering of these worlds, so that the later is the more determinate. It is to be noted that these are spatiotemporal worlds, not time-slices of worlds.

In possible world theory, a world w corresponds to a set Prop(w) of propositions, namely all those true of w. The possible worlds are usually taken to be determinate, hence the sets are usually taken to be maximal consistent sets of propositions. In my proposed actual world account, however, Prop(w), though consistent, will be sub-maximal, for all but at most one actual world. (The one exception might be a final actual world, $a_{+\infty}$ such that Prop($a_{+\infty}$) consists of all truths). The worlds are ordered by determinacy in such a way that world v is less determinate than world w just in case Prop(v) is a proper subset of Prop(w). The following formal constraints must hold:

Closure: If w is any actual world, then any propositions entailed by those in Prop(w) are also in Prop(w).

Interworld Consistency: If v and w are actual worlds, then Prop(v) $\cup$ Prop(w) is consistent.

Identity of Indiscernibles: If Prop(v) = Prop(w), then v = w.

In addition it is plausible that for any two actual worlds v and w there are actual worlds y and z such that Prop(y) = Prop(v) $\cap$ Prop(w) and Prop(z) = Prop(v) $\cup$ Prop(w).[21]

Given this ontology of actual worlds, we characterise the temporal ordering as precisely the ordering of increasing determinacy. The interworld consistency requirement then tells us that there cannot be inconsistent propositions actual at different times, and hence we can alter neither the past nor the future in the sense of unmaking the truth.

The question now arises as to what the connection is between the temporal ordering and the temporal dimension. For a given point *t* on the temporal dimension we can consider Ab(*t*), the set of true propositions about events occurring before *t*. Assuming that past actuality is wholly determinate but the future is not, then a point, T[w], on the temporal dimension will, in most cases, be correlated with an actual world w. It is the latest moment for which the history of the universe up to that moment is true at the world w. This I take to be the *present moment* in world w. I explicate this, more formally, as the greatest member (strictly speaking *least upper bound*) of the set of the *t* for which Prop(w) contains Ab(*t*).[22] Notice, however, that I am not assuming that T[w] is well defined for all actual worlds w. As we shall see, there are special cases in which no point on the temporal dimension is the present for the actual world in question.

I have first distinguished and then correlated the temporal ordering (which may be identified with the partially ordered set of actual worlds) and the temporal dimension. This is all that is required to make sense of our passage through (or more accurately growth in) time. It is, on my account, a contingent truth that the time (measured along the temporal dimension) correlated with the actual world increases as we go to later, that is, less determinate, worlds.[23] For time might have marked time! That would have occurred if the increase in determinacy had merely been an increase in distant future determinacy without any uniform transition from indeterminacy to determinacy at the point of time in question.

Something else which follows from the distinction between the temporal ordering and the temporal dimension is that we can make sense of what we might be tempted to call a 'time before time' and a 'time after time'. To avoid confusion I shall call them *alpha* and *omega* respectively. Alpha, or strictly, an alpha, is a time not on the temporal dimension but earlier than any point on the temporal dimension. Alpha would be correlated with an actual world, $a_{-\infty}$, less determinate than any w for which T[w] is defined. In this case, T[$a_{-\infty}$] fails to be well defined because Prop($a_{-\infty}$) does not contain any Ab(t). Again there could be an actual world $a_{+\infty}$, more determinate than any world w for which T[w] is defined. That would be omega. In this case, T[$a_{+\infty}$] fails to be well defined because the set of t for which Prop($a_{+\infty}$) contains Ab(t) has no upper bound. We might further speculate that Prop($a_{+\infty}$) is a maximal consistent set, so $a_{+\infty}$ is the one and only determinate actual world. Far from being a novelty omega would then be the only one of my actual worlds which would be recognised as a world by Lewis — in his account it would be *the* actual world.

The above, although technical in its details, is nothing more than a formalisation of the simple enough thesis that the passage of time is the inexorable increase in determinacy of the otherwise Parmenidean four dimensional actual worlds. If my aim was merely to explain the intuitively appealing idea of the passage of time it might not be worth it. For isn't the plurality of actual worlds just an explication and so no explanation of the passage of time? That would be a fair enough comment. But my aim is to show that given this explication of the passage of time we can now, at no further cost, explain both physical and metaphysical necessity.

I have already submitted that what is necessary at a given moment of time is precisely what is actual at that moment. So what is actual at all times will be an important sort of necessity. Or rather, since this is ambiguous, we have two important sorts of necessity: (i) that which is actual at all times of the temporal dimension; (ii) that which is actual at all times in the temporal ordering. Because there might well be an alpha world (or several alpha worlds) these two necessities need not coincide, in which case the first is weaker than the second. I take the first to be physical necessity and the second to be metaphysical necessity. The set of propositions which are physically necessary is $\bigcap$\{Prop(w): for some t, t = T[w]\}. Putting it another way, something is physically necessary if it is part of every world for which there is correlated a present time on the temporal dimension.

The set of propositions which are metaphysically necessary is $\bigcap$\{Prop(w): w is an actual world\}. That is, something is metaphysically necessary if there is no actual world in which it fails to occur. Suppose, for simplicity, that there is just one alpha world, $a_{-\infty}$. Then, because the temporal ordering corresponds to the ordering of sets of proposition by inclusion, $\bigcap$\{Prop(w): w is an actual world\} = Prop($a_{-\infty}$). So the metaphysically necessary truths are just those true of the highly indeterminate first actual world. Now $\bigcap$\{Prop(w): for some t, t = T[w]\} $\supseteq$ Prop($a_{-\infty}$), but even though there is no world w correlated with a point on the temporal dimension for which Prop(w) = Prop($a_{-\infty}$), the formalism leaves it quite open as to whether $\bigcap$\{Prop(w): for some t, t = T[w]\} = Prop($a_{-\infty}$) or $\bigcap$\{Prop(w): for some t, t = T[w]\} $\supset$ Prop($a_{-\infty}$).

So the formalism leaves it quite open as to whether the physically necessary and the metaphysically necessary coincide. I may, therefore, distinguish the two by insisting that $\bigcap$\{Prop(w): for some t, t = T[w]\} $\supset$ Prop($a_{-\infty}$).

On the theistic interpretation of metaphysical necessity we should take $a_{-\infty}$, the alpha world, to be the way the universe was prior to creation. So, I say, $a_{-\infty}$ is part of God, because the situation of total physical indeterminacy is uncreated and whatever is uncreated is part of God. Atheists should say that $a_{-\infty}$ is the way the universe was before the *as if* creation which resulted in the Big Bang. That amounts, though, to nothing more than saying that $a_{-\infty}$ is the way the spatiotemporal universe was at the moment of the temporal *ordering* earlier than any moment on the temporal dimension.

My aim has been to give an alternative explanation of physical necessity, not an alternative description of what is physically necessary. It is worth pointing out, however, that I have a solution to the uninstantiated kinds problem. First, I should say something about mere possibilities, including uninstantiated kinds (and other uninstantiated universals). The question: 'Do they exist in the actual world which is correlated with the present, or with any future time?' receives the answer 'No'. But there is at least one past (i.e. less determinate) actual world in which that question has neither a 'Yes' or a 'No' answer. In that way the merely possible depends on the actual. Having given this account of uninstantiated kinds, the way I characterized the physically necessary and the metaphysically necessary suffices to distinguish those which are physically possible from those which are physically impossible, but metaphysically possible. If we are able to conceive of kinds which are metaphysically impossible, then they would be ones for which the question of their existence receives a 'No' answer in all actual worlds.

5. CONCLUDING REMARKS

I have posited a plurality of indeterminate actual worlds in order to explicate both the intuitive datum that in some sense time passes and to explain both physical and metaphysical necessity. Alternatively, we could think of the actual worlds as a heuristic model, much as many consider possible worlds to be. In that case it is the nature of time, reflected by this modelling, which explains physical and metaphysical necessity.

But is it worth it? we should ask. In support of my explanation of physical necessity I first, of course, remind readers of the problems with the rival accounts which I have mentioned. In addition, I shall argue that a plurality of actual worlds is in fact a frugal ontology. If worlds are taken to be aggregates of entities, then all the many actual worlds are parts of the aggregate of all such entities, which aggregate is the one determinate actual world $a_{+\infty}$.

Realism about the plurality of actual worlds amounts, therefore, only to granting that of some but not other, parts of $a_{+\infty}$ are themselves actual worlds. Now there is a strong case for the objective naturalness of some but not other classes, and we might or might not seek to find some further ontological ground for that naturalness.[24] Likewise that some but not other parts of $a_{+\infty}$ are actual worlds could be taken as a primitive fact, or some further account might be given.

University of New England, Armidale, New South Wales

NOTES

* I would like to acknowledge some extremely useful comments from David Armstrong, from Brian Ellis and from the anonymous referee.

1 My account has some similarities to the account of physical necessity in terms of branching universes given by Storrs McCall in his (1994).

2 See Forrest (1993) for more detail.

3 Cartwright (1983); Ellis (1987) pp.52-6.

4 Harré and Madden (1975).

5 Dretske (1977); Tooley (1977); Armstrong (1983).

6 Ellis (1990) pp.84-5; Ellis and Lierse (1993).

7 Bigelow and Pargetter (1990) Ch. 5.

8 Bas van Fraassen in his (1989) denies that there is a modality of physical necessity.

9 Forrest (1986); Bigelow and Pargetter (1990) pp.203-13.

10 Realists about the non-actual may characterise modalities without circularity provided the only non-actualities they believe in are ones they are content subsequently to call metaphysically possible. Thus they should insist that while talking donkeys are real non-actualities, round squares are not. Talk of realism about *possibilities* tends to obscure the non-circularity of the characterisation. For recent discussion of this issue see Miller (1989); Lycan (1991); and Miller (1993).

11 Harré and Madden (1975) Ch 6.

12 This explanation holds only given realism about non-actualities. For that reason, presumably, Armstrong does not propose this explanation.

[13] Ellis (1990) p.84.
[14] This is a version of the Andersonian doctrine of the infinite complexity of things. I find this attractive but, like Armstrong, I think of it as merely an epistemic possibility, not something we have warrant to believe.
[15] Armstrong (1983) pp.143-46.
[16] Quantum Theory may perhaps involve some backwards causation, but not, usually at the macroscopic level. See Forrest (1985) and Forrest (1988) p.133.
[17] Lewis (1986) pp.198-209.
[18] Williams (1951) pp.463-64.
[19] The snowball theory, according to which all the past is real but the future real only in so far as it is determined by the past, is considered central to Bergson's philosophy (Kolakowski (1985) pp.2-3). To my mind, however, this is obscured by Bergson's idealist tendencies, according to which time is imposed by the mind.
[20] Armstrong (1987).
[21] In jargon: the world y is then the *meet*, $v \wedge w$, of v and w, the world z is the *join*, $v \vee w$, of v and w; the actual worlds form a lattice; and the function λ_xProp(x) is a one to one homomorphism from the lattice of actual worlds into the lattice of sets of propositions.
[22] If, contrary to the common sense view, the past actuality is not fully determinate, then there will still be a, perhaps imprecise, boundary between values of *t* for which Prop(w) contains most of $Ab_m(t)$ and values of t where it does not. Here, $Ab_m(t)$ is the set of true propositions about 'macroscopic' events, that is those which a suitably placed human observer could discover by ordinary means of observation.
[23] In jargon: it is contingent that the function $\lambda_x T[x]$ is not merely monotone non-decreasing but in fact monotone increasing.
[24] Lewis (1986) pp.59-61.

REFERENCES

Armstrong, D. M., (1983) *What is a Law of Nature?* Cambridge: Cambridge University Press.

— , (1987) 'Smart and Secondary Qualities' in Pettit, P., Sylvan, R. and Norman, J. (eds.), *Metaphysics and Morality: Essays in Honour of J. J. C. Smart.* Oxford: Blackwell.

Bigelow, J. C. and Pargetter, R. (1990) *Science and Necessity.* Cambridge: Cambridge University Press.

Bigelow, J. C., Ellis, B. D. and Lierse, C. (1992) 'The World as one of a Kind: Natural Necessity and Laws of Nature', *British Journal for the Philosophy of Science* 43, 371–88.

Cartwright, N. (1983) *How the Laws of Physics Lie.* Oxford: Oxford University Press.

Dretske, F. I. (1977) 'Laws of Nature', *Philosophy of Science* 44, 248-68.

Ellis, B. D. (1987) 'The Ontology of Scientific Realism' in Pettit, P., Sylvan, R. and Norman, J. (eds.), *Metaphysics and Morality: Essays in Honour of J. J. C. Smart.* Oxford: Blackwell.

Ellis, B. D. (1990) *Truth and Objectivity.* Oxford: Blackwell.

Ellis, B. D. and Lierse, C. (1993) 'The Fundamental Importance of Natural Kinds'. Paper read to Australasian Association of Philosophy Conference, Adelaide, July, 1993. Available as *Victorian Centre for the History and Philosophy of Science Preprint, 3/94.*

Forrest, P. (1985) 'Backward Causation in Defence of Free Will', *Mind* 94, 210-17.

— (1986) 'Ways Worlds Could Be', *Australasian Journal of Philosophy* 64, 15-24.

— (1988) *Quantum Metaphysics.* Oxford: Blackwell.

— (1993) 'Just Like Quarks? The Status of Repeatables' in Bacon, J., Campbell, K. and Reinhardt, L. (eds.), *Ontology, Causality and Mind: Essays in Honour of D. M. Armstrong.* Cambridge: Cambridge University Press.

Harré, R. and Madden, E.H. (1975) *Causal Powers: A Theory of Natural Necessity.* Oxford: Blackwell & Mott.

Kolakowski, L. (1985) *Bergson.* Oxford: Oxford University Press

Lewis, D. (1986) *On the Plurality of Worlds.* Oxford: Blackwell.

Lycan, W. G. (1991) 'Pot Bites Kettle: A Reply to Miller', *Australasian Journal of Philosophy* 69, 212-3.
McCall, S. (1994) *A Model of the Universe.* Oxford: Oxford University Press.
Miller, R. B. (1989) 'Dog Bites Man: A Defence of Modal Realism', *Australasian Journal of Philosophy* 67, 476-78.
— (1993) 'Genuine Modal Realism: Still the Only Non-circular game in Town', *Australasian Journal of Philosophy* 69, 159-60.
Tooley, M. (1977) 'The Nature of Laws', *Canadian Journal of Philosophy* 7, 667-98.
van Fraassen, B. C. (1989) *Laws and Symmetry.* Oxford: Clarendon Press.
Williams, D.C. (1951) 'The Myth of Passage', *Journal of Philosophy* 48, 457-72. Reprinted in Gale, R. M. (ed.) (1978) *The Philosophy of Time: A Collection of Essays.* New Jersey: Humanities Press.

F. JOHN CLENDINNEN

THEORIZING AND EMPIRICAL BELIEF *

It is generally held that induction contributes little to the appraisal of competing theories. Certainly our expectation that a theory which has adequately explained observed facts of a certain kind will continue to explain facts of the same kind may be considered to be inductive, in a broad sense of the term. However if, as is widely held, theories are underdetermined by the empirical facts for which they account, there must be other criteria which play a major role in selecting between competing theories. In recent decades the historical variability and the social dependence of the criteria that usually are applied has been emphasized and this presents a problem about the rationality of theory selection. Larry Laudan has recently offered an account of theory appraisal in which induction plays a much more substantial role. Specific methodological rules are accepted according to their past success in selecting theories that exhibit the epistemic virtues we seek in them (Laudan 1987). According to this account, criteria of appraisal do change but change rationally. Induction plays a crucial role, yet it is only external; it selects the criteria we employ, but these criteria are in no way implicit in induction itself. I will argue that induction actually contributes much more than this. When an adequate specification of induction is provided (a task that is not as simple as is sometimes assumed) it is found to include criteria which play a substantial role in the appraisal of theory.

As far back as 1953, Nelson Goodman pointed to a serious problem about induction which has never been satisfactorily resolved. (The lectures in which he did this are published in Goodman (1965)). He pointed out forcefully that induction involves more than accepting those hypotheses which are compatible with all empirical data and which have, or would have had if used, a history of predictive success. However the solution Goodman suggested implied that rational standards are the outcome of historical contingencies. Others suggested a different solution which I believe to be correct, but which has not gained general support and needs to be integrated into a broader account.

I will argue that a solution of Goodman's riddle should be implicit in an adequate specification of the empirical method which includes induction, the *prima facie* acceptance of perceptual beliefs and the criteria by which we appraise these beliefs. To establish that a proposed specification of this method is adequate we will need to show that the policy as specified is rational. A key point is that our most basic empirical beliefs are about what is external to us and are defeasible. I will argue that the appraisal and reinterpretation of these beliefs involve procedures which also serve as the starting point for theorizing. In

P.J. Riggs (ed.), Natural Kinds, Laws of Nature and Scientific Methodology, 63–92.

particular, the standards by which we establish which objective properties are more basic play a crucial role.

The thesis that I argue applies most directly to the part of theory which lies closest to the observable world and so may be seen as not very exciting. However I believe that clarity about this less sophisticated part of theory provides the best perspective for tackling the many challenging epistemological and ontological problems that have been thrown up by modern physics.

1. SPECIFYING INDUCTION

Goodman's point (1965, p.72 *et seq.*) can be put in its full generality thus: Any hypothesis H is equivalent to the conjunction of a pair of hypotheses, one dealing with a set of facts which includes all those already known directly and the other dealing with facts which are not. In particular we can take the first conjunct as including just those facts which are directly known in a certain context, C, and the other as including all the content which is predictive in context C. That is:

(1a) Any hypothesis, H, is equivalent to the conjunction of an evidential conjunct, H(d), and a predictive conjunct, H(p), where H is under consideration in context C, H(d) deals only with data available at C and H(p) is an hypothesis which is entirely predictive at C.

The requirement that H be compatible with known data places a constraint only on the conjunct H(d); but none of its content is predictive in the context C. Consequently the requirement that we adopt hypotheses which are compatible with known data has, taken alone, no predictive consequences. Predictions are only implied when we accept a constraint on the kind of hypotheses that may be considered. Combining an arbitrary H(p) with a given H(d) can give us an hypothesis which will lead to intuitively bizarre predictions. Indeed by a suitable choice of H(p) any predictions at all can be arrived at while abiding by the constraint that the predicting hypothesis must conform to known data. It is clear that we do exclude most of the possible combinations of data and predictive components into hypotheses because we have a notion of the appropriate relationship between them. We need to formulate this notion into a precise constraint which selects those hypotheses which are supported by evidence that conforms to them. Since a predictive hypothesis asserts a pattern in the facts which projects beyond what at that time is established by observation, adopting such an hypothesis can be described as a *projection*. Goodman accordingly speaks of determining which hypotheses are valid projections, or, more briefly, which hypotheses are projectible (Goodman, 1965, p.84).

For Goodman the crucial question was what predicates may legitimately be used in inductive hypotheses. This is indeed important, but we should note that it is by no means enough to avoid bizarre predictions. The hypothesis:

(1b) All emeralds are green until T and thereafter blue

uses only standard predicates of English and is compatible with all we have observed up to T. Yet using it at any time up to T would lead us to predict that any emerald will be blue after T; a conclusion at odds with the prediction we take to be inductively supported. Goodman is assuming that we accept a suitable syntactical constraint in using induction and that this constraint precludes hypotheses such as (1b) as syntactically illegitimate. The point he wished to stress was that such syntactical constraints could always be circumvented by adopting predicates suitably defined in terms of those we ordinarily employ. In particular, consider the predicate 'grue', which may be defined thus:

(1c) x is grue at t iff t is not later than T and x is green, or t is later than T and x is blue.

(This is not exactly the definition Goodman proposed but it makes his point just as well and Goodman has gone along with this definition in discussion.) Using the predicate so defined we can formulate a sentence,

(1d) All emeralds are always grue,

which is of the same form as 'All emeralds are always green' and yet means exactly the same as the syntactically complex hypothesis (1b). A theory of projectible hypotheses must have, as a minimum, a formal requirement on the relationship between the hypothesis and the body of known facts (Goodman 1965, p.72). In addition there must be some non-syntactical constraint on the hypothesis (*ibid.*, pp.72-3, 83). So he proposes a criterion to mark off those predicates which must be used if an hypothesis is to be projectible, or rather which determines the degree of projectibility. This criterion is the degree of comparative entrenchment of the predicates employed (*ibid.*, p.108). He suggests that this is the non-syntactical determinant of the projectibility of hypotheses.

This theory has a grave defect. The degree of entrenchment of a predicate is determined simply by the use that has been made of the predicate in previously projected hypotheses. It follows that according to the theory there is nothing about those predicates that have been used which makes them superior in any way to other predicates that we might have used. Goodman writes:

> [T]he judgement of the projectibility has derived from the habitual projection, rather than the habitual projection from the judgement of projectibility (*ibid.*, p.98).

The requirement that past habitual projections should have been successful is no help. If we had habitually used 'grue', defined with T in the future, then the hypotheses formulated would have had exactly the same degree of success as we have had up to now using 'green', but it would have led us to predict that emeralds will be blue. Goodman agrees that we should not do this, but only because of what our habits happen to have been. We cannot claim that the policy we have adopted is any more rational than a policy that we could have adopted and would have led to conflicting predictions. It is hard to see how anyone who grants this can go on to claim that either of the competing predictions is more reasonable than the other. To do so would make reason dependent on mere habits. (It would be incompatible with Goodman's position to hold that the habits could have been appraised, even although they were not; for this would imply the presence of some feature of the procedure which, if it had been detected, would have shown it to be justified. Then it would have been this feature and not entrenchment, that established projectibility.)

Goodman adopts his entrenchment theory of projection arguing the lack of a satisfactory alternative. It may seem that (1b) (i.e. that emeralds are green until now and hereafter blue) can be rejected because it asserts a pattern in the future which differs from what has occurred until now. However Goodman's point is that although the pattern changes relative to a characterization in terms of 'green' and 'blue', it is the same relative to 'grue'. He acknowledges that some will hold that this latter predicate is not 'purely qualitative', in that it implicitly refers to a position in time. However he dismisses this suggestion and holds that the requirement that 'positional' predicates be not used in induction turns out to be vacuous. While the definition of 'grue' in terms of 'green' and 'blue' includes reference to a position in time (as does the definition of 'bleen' by ' "x is bleen at t" iff "t is not later than T and x is blue, or x is later than T and x is green" '), so does the definition of 'green' in terms of 'grue' and 'bleen'. The two pairs of definitions are exactly symmetrical (*ibid.*, pp.79-80).

What the symmetry of definitions shows is that appeal to the definition alone is not sufficient to establish positionality. However it does not establish that 'grue' and 'bleen' are not positional by some criterion other than definition. If ordinary predicates are non-positional and can be established to be so without invoking a definition, then the time reference in the definitions of 'grue' and 'bleen' does establishes them as positional. (If these predicates should then be used as *definiens* in so called definitions of 'green' and 'blue', their positionality is cancelled out by the positionality of the definition. So we end up with non-positional *definienda*. That this may be so is plausible; for we start off knowing the meaning of 'green' and 'blue' but can only understand 'grue' and 'bleen' when Goodman provides definitions.)

1.1. *Positionality and Psychological Nominalism*

According to Goodman the notion of positionality, if introduced, would be entirely relative to our linguistic practice; which predicates we should use (and which would be non-positional) depends only on which we start with. This suggests the thesis of Psychological Nominalism, discussed, for instance, by Wilfrid Sellars. He characterizes it as denying that there is any awareness of kinds, or even of resemblance 'prior to, or independent of, the acquisition of language' (Sellars 1956, pp.289, 291). This challenges the seductive assumption that the boundaries of the classifications we apply to entities in the objective world are simply there for us to experience directly. However it goes much further than this, for it denies that the recognition of resemblances can precede and be a condition for, learning a language.

While sympathetic to psychological nominalism Sellars makes some very plausible concessions that seem to undercut it. He does not deny that there are objective similarities and differences and that these are causes of the way we associate words with objects of a certain objective kind. He further allows that this causal connection is mediated by our sensations (*ibid.*, p.290). If it is granted that we are aware of our sensations, it is hard to see how we can be unaware of the resemblances. Indeed if we are not initially aware of these resemblances it is hard to see how we could recognize that a new item should be included in the extension of a predicate. That is, it would be a mystery how we could learn to apply a predicate in general by learning that it was accepted practice to apply it to a finite set of individuals. It is perhaps possible in principle that we might simply find ourselves using a predicate in a certain way, without any understanding of what we do. But in fact we certainly are aware of the resemblances in appearance in virtue of which we apply a colour predicate to an object observed for the first time.

If there were no prelinguistic resemblances and Psychological Nominalism were correct, the extensions of the predicates we use would be entirely a matter of our choice; and this would sustain Goodman's claim that 'grue' is no more positional than 'green'. But we have seen the difficulties with this thesis (in particular the difficulty of making sense of a choice of extension which could be in any sense systematic). In any case, at one point Goodman allows that there are perceptual resemblances. He says '"Grue" might be classed as positional on the ground that its instances of application before [T] cannot match those after [T] in colour, while this does not hold for "green"' (Goodman 1968, p.163, n.1).[1] He tells us that this had been pointed out to him by Noam Chomsky but dismisses it as 'too *ad hoc* to be of much interest'. He points out that this criterion will not exclude all predicates that are non-projectible and will be inapplicable to theoretical, non-observational predicates. Thus he allows that we can recognize a degrees of resemblance between all instances of 'green', even if they are often discriminable in shade from each other. Also, he allows that these resemblances are independent of the predicates we use, since instances of the same predicate 'grue' will not match. By implication this is a rejection of psychological nominalism as defined by Sellars.

So Sellars makes concessions that undermine psychological nominalism. Goodman effectively rejects this doctrine and it conflicts with our experience in using descriptive predicates like those of colour. I conclude that we can recognize many resemblances (certainly those of colour) prior to learning to use the relevant predicates. Since Goodman agrees, let us consider his reasons for denying that recognized resemblances between the instances of a predicate might determine that it is projectible. His reasons are that this criterion is too *ad hoc* and that there are predicates which are not projectible but which are not excluded by this criterion. These two reasons are closely linked. An account of which predicates are projectible must be applicable to predicates which do not designate observable properties. As Barker and Achinstein had pointed out, grue-like theoretical predicates can be constructed and the use of these would affect theoretical projection just as the use of 'grue' affects everyday inductions. If all projectible predicates were observational and involved recognizable resemblance throughout their extension, then resemblance of instances might be held to be what determines projectibility. But it is not; and so simply invoking lack of resemblance in a particular case as grounds for not projecting can fairly be described as *ad hoc*. Goodman is making a valid point, but it establishes much less than he suggests. All he has done is point out that we can not assume that perceptible resemblances are relevant until we have a theory of projection which explains why it is, at least in certain cases. Wesley Salmon proposed such a theory: that in primitive inductive inferences only predicates based on perceptible resemblance may be employed (Salmon 1963, pp.38-40). This theory does not hold that all and only predicates picking out perceptible resemblance are projectible, but it does imply that 'green' is projectible in inductions about coloured objects. I do not believe that it is possible to draw a clear line between those inductions which are and those which are not, primitive. So I believe that Salmon's theory needs some modification; however I will argue in what follows for an account not very different from his.

2. INDUCTION AND THE EMPIRICAL METHOD

A theory of projectible hypotheses should emerge from a satisfactory specification of induction; and the adequacy of a proposed specification should be established by showing that the method it specifies is the rational way of predicting. To make decisions about actions we must predict. So making predictions is crucial to an activity basic to humans. We have all sorts of wants, concerns and fears. We think in order to guide actions aimed at achieving some states and avoiding others. So let us try to show that induction is the rational way of using thought in this way. Such a justification of induction will be instrumental and so relative to what is sought. Nevertheless it has a certain universality: it is not relative to any specific goal but only to the having of goals. So it will carry weight for anyone who wants to do anything at all and that is a wide enough category to give the justification point. Nevertheless we may hope for more: that induction will lead to truth. I suggest that initially we see how far we can get with a justification relative to our practical concerns; for what this achieves may illuminate what we do in seeking truth.

Induction cannot be isolated from the method by which we appraise our specific empirical beliefs. Firstly inductive inference presupposes some initial beliefs and its conclusions can be no more secure that the premises. Secondly any belief about the objective world has predictive content, so adopting such beliefs is already a kind of induction. Finally induction must play a substantial role in establishing the trustworthiness of even the most direct, experientially based objective belief. There is something of a circle in this last point, but it is not vicious. I will argue that we can recognize the rationality of induction when it is characterized in only the most abstract terms — terms which are implicit in the very notion of prediction. Then, armed with this minimal standard of rationality, we can turn to the rationality of accepting beliefs about a putatively objective world and consider the standards we should apply in adopting such beliefs.

The very idea of practical action presupposes criteria of success and failure. Equally the notion of predicting implies that there is a question about its correctness, a question that must be settled by something objective. The predictions we actually make are typically about the concrete world and so about what is paradigmatically objective. Physicalists claim all significant beliefs are, implicitly if not explicitly, of this kind. I do not assume physicalism (indeed I am dubious about it) but would insist that whatever we can significantly speak of predicting about, if not physical must nevertheless be objective in the sense that it is what it is irrespective of the wishes and beliefs of people. I will argue that if we take 'prediction' in the abstract sense of the formation of a belief which will subsequently be objectively confirmed or refuted, we have enough to establish induction — again characterized in abstract terms — as the rational way of predicting.

I will draw on a justification which I proposed some time ago (Clendinnen 1982, 1986) which was very much in the spirit of the work of Hans Reichenbach and Wesley Salmon, but which differed from that approach in some important respects.[2] In this earlier work, I treated induction in isolation from the acceptance of perceptual beliefs, an approach which I now see as misleading. However I still see much of the earlier work as valid and in this essay will pass in outline over arguments I have already developed in more detail. Since I want to initially deal with induction under as abstract and general a characterization as possible, I will try to avoid any assumptions about the kind of things it is about; and this means avoiding any commitment to a specific object language. Having argued for the rationality of induction so conceived, I will then move on to the standards for appraising and generalizing beliefs about the concrete world; arguing that these standards emerge from our more general conception of induction. I will be particularly concerned to show how these standards determine which property-concepts should be employed in induction, and also the standards which should guide our theorizing about those parts of the world which only affect our sensory apparatus indirectly. But first, the justification of induction.

2.1. *Predicting without Guessing*

It is not possible to establish that any procedure is either necessary or sufficient for a prediction to be successful; so the most obvious kinds of instrumental arguments are excluded. What remains possible is to compare induction to alternative policies to see whether it is superior in any respect. When we do, I believe we can show that all possible alternatives are irrational. Since it is clear that inductive predictions can succeed, and in lieu of any argument that induction is irrational in some way other than that in which its competitors are, we are left with it as the one rational way of trying to predict. The strategy I am proposing implies that there are standards of irrationality which are logically prior to positive standards of rationality. This may seem paradoxical, but there is no doubt that we can sometimes recognize a certain approach as self-defeating — and so irrational — while lacking any positive grounds for expecting any other approach to succeed. Typically, in such cases, we need to call on general empirical knowledge, however sometimes the very nature of a policy ensures that it is self defeating. So it is sometimes possible to establish the irrationality of a policy by conceptual analysis and without invoking any standards which determine what policy would be rational. This is the key to establishing the rationality of induction and, indeed, of all the principles that constitute the basic empirical method.

If we explore the possible alternatives to induction, we are able to see that they all must involve, at some stage, a pure guess. (In speaking of a pure guess I intend to exclude those cases in which one accepts what comes first to mind in a certain context, but where there is a record of beliefs so formed being later shown correct. Such cases might be called 'guesses' but there are facts, other than one's inclinations, which are relevant to trusting this policy. That we act rationally in such cases must wait on induction being shown to be rational. At this stage all that is necessary is to distinguish these cases from pure guesses, where no collateral, factual considerations are involved.) While pure guesses do not necessarily fail — one may always turn out true by good luck — relying on a guess drastically limits the use of thought in guiding our expectations. The outcome of an intellectual procedure that includes a step which is a pure guess is at the mercy of what is neither an objective input or a determinate procedure. Using the intellect in this way is self-negating. The point about a guess is that it is not subject to appraisal. If what we end with depends on the guess, any thought that went before it is made pointless; and if one tries to review a pure guess, one thereby becomes aware of the alternative conclusions that might have been adopted, and that they are all on a par. If there should be some way of selecting between them, then what was selected would not have been a pure guess. So an attempt to appraise a pure guess leaves us with a whole array of possible outcomes. If we leave it at that, we have no prediction. If we nevertheless settle on one alternative we act on nothing more than an inclination, an inclination which is not necessarily stable. The cost of settling on a particular prediction is that thought is thereby cut short; if we persist in thinking about what we are doing the alternatives are back before us and we have failed to discriminate between them.

2.2. *Projecting Past Patterns*

People often accept without thought their inclinations to belief and it would clearly be ridiculous to suggest that we should critically scrutinize all the beliefs we entertain. However if thought has any role in guiding what we do, all beliefs must be open to critical scrutiny. If we question a belief and then simply dismiss the question, we effectively deny that thought has any role to play. Likewise if we simply follow our inclination to a belief. To suggest that following an inclination is grounding a belief is to confuse a subjective tendency with something to which thought can appeal. In the latter case, the appeal must be to something objective in the sense that its status is independent of our thinking: something that is given to thought. It is of course, a matter of fact that one was inclined to a certain belief at a certain time. May this not be a relevant datum in considering acceptability? Maybe; we might have inductive evidence that beliefs that come to mind in such a context usually turn out true. If we hold that the fact of an inclination is evidence for a prediction, we must face a question as to why it should be so taken. However if there is no answer, we are following an inclination and not using the inclination as evidence.

Having established the irrationality of trusting pure guesses, I contend that we can show that any deviation from induction must involve just this kind of irrationality. In outline the argument, which is developed more fully in my (1982), is this. To predict without guessing we must follow a systematic policy; and doing so involves expecting a pattern, such as an E-type item of data always being followed by a P-type item. If we now ask ourselves whether there is any way that the data presently available could most directly, and without arbitrary choice, indicate what pattern will be present in what we will receive as data, the answer is that it can if we take patterns which persist through all we know by experience to persist. We cannot be neutral about these patterns because we want to predict and if our predictions succeed there will be a regular pattern in future facts. The question must then arise as to whether future patterns will be the same or different from what has been. Sometimes we may expect a change in pattern on the basis of what has happened. This will be so if this change is involved in the persistence of a larger, already detected, pattern, of which the first is a part. But here we still take the most comprehensive pattern to persist.

To expect that the most comprehensive pattern we can detect will change would commit us to guessing, to guessing the occasion of the change and the kind of change. Of course it is possible to take some available datum other than the persistence of the pattern that has occurred as indicating a future pattern. Since our expectations would then be guided by what is given objectively according to a rule, the prediction would not be just a guess; but a pure guess would nevertheless be involved. We would have to consider why the datum in question should be taken as indicating a specific pattern. If it is not a guess then some other datum must be invoked and then its relevance is in question. We have started on a regress which can only end by just taking some datum as relevant in a certain way and doing this is guessing. The conclusion is that to expect the most comprehensive pattern we can detect to change must be an arbitrary guess. On the other hand taking it to persist enables us to make predictions using available data in the most direct way possible. We can give up trying to anticipate what will follow what, or we can use induction, or we can guess or build some policy on a guess. Those are the options. Neither reason nor anything else can assure us that any prediction will succeed. However we can identify the possible courses open to us and identify which are irrational. This points to the following policy of predicting as the only one which is not disqualified as irrational:

(2a) Identify a pattern which persists throughout all the data and predict in a way which will be correct if that pattern persists.

What we have so far is a significant constraint on rational predicting. However the requirement that we project the same patterns which persist through the available data remains vague until we have a clear notion of what constitutes a pattern. Goodman has shown us that this is no trivial matter. There are many ways in which a pattern can be compounded out of parts. Characterizing an objective sequence of facts via certain concepts may imply that there is no change, while employing a different system of concepts may imply that there is. If we do not impose some constraint on the *property-concepts* we employ, an expectation that patterns will remain the same is vacuous; for whether or not there is a change of pattern is relative to the concepts which are taken as giving the pattern its identity.

3. THE DATA OF INDUCTION

The patterns we are concerned with are constituted by facts, about which we acquire empirical beliefs which serve as premises for inductive reasoning. So an examination of our understanding of the patterns we project must start with an examination of these beliefs. This must include a consideration of their grounding, for, as I have urged, it is only by establishing that a method under a certain specification is rational that we can be sure that the specification is satisfactory.

3.1. *Foundations and Coherence*

Contemporary epistemology has largely directed attention away from the status of empirical data. Some decades ago, those who took this question seriously generally assumed that basic empirical beliefs were about our subjective states, so that all material-object beliefs were inferred. In spite of the profound difficulties with this thesis, its advocates were pushed towards it because they assumed that basic beliefs must be incorrigible and material object beliefs obviously are not. It has only been more recently that some epistemologists allowed that there are non-inferred empirical beliefs which are nevertheless defeasible. This makes it possible to hold that our most basic beliefs are themselves about the world. Beliefs about the world which are based on perception may be taken to have a *prima facie* but defeasible warrant. These beliefs can then serve as the starting point for generalized beliefs which provide both a basis for prediction and also, by systematizing individual beliefs, provide a means for appraising the perceptual beliefs which were initially accepted *prima facie*. Susan Haack, in her recent book (1993), gives a full and detailed account of this kind of epistemology, which she calls 'foundherentism' (because it draws on both foundationalist and coherentist epistemologies). I hope that this essay will contribute to the reconstruction she seeks to initiate in epistemology.[3]

Amongst those who had earlier moved in a similar direction is Wilfrid Sellars, who in his (1973) and (1979), clarified his attitude to foundationalism. In spite of his (1959) refutation of the classical version of that doctrine, he insists that there is a set of basic empirical beliefs which are *prima facie* warranted just because of the kind of belief they are. I propose an argument aimed at showing why it is reasonable to accept perceptual beliefs as having a *prima facie* warrant which I believe follows very closely Sellars' (1979) argument; however others have interpreted him differently (Triplett 1990, p.95, cf. *ibid.*, pp.98, 100, 101). Since it is not my concern here to argue a question of interpretation, I will note my debt and develop the justification in my own way, drawing on the idea that is central to the foregoing justification of induction. Here, in outline, is the argument.

Given that rational prediction is by induction, we must seek patterns among the data of experience. It is this that obliges us to put *prima facie* trust in our perceptually based beliefs about the objective world; for it is only by taking experience as originating from an objective world that we can identify regular patterns. There are objective interpretations of experience that come to us naturally and if we do not put some trust in these beliefs we are either left with no patterns to project, or must guess what hidden patterns there might be. There can, of course, be no assurance that perceptual beliefs are all correct; indeed we have good reason to believe that quite a few are not. So our trust in them must be qualified. Nevertheless that a degree of *prima facie* trust is appropriate is reinforced when we consider how we come to recognize that some of these beliefs are erroneous: it is, after all, on the basis of empirical evidence. Thus it is by placing trust in some perceptions that we come to correct others. If we ask how we come to trust some perceptions more than others, the answer is that it is by appeal to our generalized systematization of belief. All perceptual beliefs have some claim on us, but some need to be abandoned or modified if our system of beliefs is to be maximally integrated under regularities. The procedure involves a balance of the claims of perception against those of unification. While it is not possible to formulate an algorithm for striking this balance, there are criteria that guide it. These criteria, which are needed to guide our crudest generalizations about what is observable, are sufficient to guide theoretical speculations about structures and processes which can not be directly observed.

3.2. *Perception and Interpretation*

If we did not take experience as originating from an objective world, we would be obliged to seek patterns in the experience itself. The claim that we recognize persisting patterns in items of experience as such implies that our objective concepts, for instance of solid three dimensional objects, are logical constructs, from items of subjective experience, which are employed because they are convenient, but are actually unnecessary for prediction. This is phenomenalism. The difficulties with this doctrine are well known and epistemologists take care to deny it any kind of foothold. This has sometimes led to denials that we are aware of the items of subjective experience which cause our basic material-object beliefs. However this is too strong. Certainly, as a matter of psychological fact, our beliefs about concrete objects are typically not mediated by more basic beliefs about subjective experience. Yet we have no doubt that these beliefs are caused by the stimulation of sensory nerves. Nor is there any doubt that we are sometimes aware of the subjective states caused by sensory stimulation. Indeed even with sensory modes like sight and touch, which normally lead directly to objective beliefs, it is possible to focus on the subjective states. In rare unfamiliar contexts we are actually forced to do so and to deliberately consider what structure of objects might have caused it. (Consider structures deliberately designed by psychologists, for pedagogical and experimental reason, to contain conflicting perceptual cues.)

I will argue that it is important to acknowledge that perceptually based beliefs are interpretations of something given in a more basic sense. The complexity of the process involved in the recognition of patterns and of three-dimensional objects, is attested by the degree of attention that such topics receive in cognitive science. However, the point can be made without going beyond our every day knowledge of perception. We are often obliged to revise an observation-based belief while allowing that what was given in sensory experience was not distorted. There are, by contrast, cases where there is distortion, where the causal mechanisms of perception are in some way overridden: cases of hallucination. (There was nothing about Lady Macbeth's hands that caused the sensation of red involved in her belief that her hands were covered in blood.) However hallucination can be contrasted to cases of illusion which work the same way for all observers in the same context. (It seems that an extraordinarily tall man walks on to a theatrical set, when he is actually of average height but close to part of the set that has been shrunken to give a perspective of depth.) Illusion highlights the fact of interpretation, for that is what goes wrong in these cases. The role of interpretation is also evident in the predictive content of perceptual belief. It is notorious that any material object belief is fallible and it is experience that does the falsifying. When I have a belief about some material object I have the capacity to recognize that some possible future experiences are compatible with that belief and that others are not. I can readily recognize this when I have visual or other experience that conflicts with my belief and I have some ability to imagine the future experiences which are necessary, in the respective contexts, if the belief is to remain unchallenged.

3.3. *The Refutation of Phenomenalism*

Some see the admission that we can be aware of perceptual experience itself, and can have beliefs about it, as a concession that makes phenomenalism irresistible. In particular, if material object beliefs are to be confirmed or negated via experiential data, then surely the full predictive content of a belief of the former kind can be equated with the totality of possible experiential beliefs which are relevant to its truth. If this were so, the claim that objective beliefs are necessary to facilitate prediction would fail. Consider (a) a set of descriptions of the experience which evidences an objective belief; (b) the statement that expresses that belief; and (c) the set of statements describing experiences which would be relevant to the truth of the material object statement. If there are rules for proceeding from (a) to (c) via (b), and if (b) is analyzable into concepts used in (a) and (c), then there should be a rule linking (a) directly to (c). An immediate problem for this analyzability thesis is that we lack a satisfactory vocabulary for directly describing subjective experience. When we are obliged to talk about experience, we often resort to saying that it is the kind of experience that typically occurs in some objectively described situation. This lack of vocabulary may well be significant; but let us be generous to the defender of phenomenalism and not press the point. Let us consider the possibility of predicting experience directly from given experience, on the assumption that an adequate vocabulary to describe experience could be developed if we took the trouble.

We must acknowledge that the predictions implicit in a belief about a material object are conditional and infinite in number. This is clear even if we consider only the visual evidence about the shape of a simple material object. For any specified view point there must be a visual experience that would be had if an observer should look from that point. There are an unlimited number of angles and distances from which the object can be viewed. Are there rules which cover this whole range of possibilities? Given a specification of the shape of the object, it is possible to calculate by geometry its projection onto a plane orthogonal to a line of sight and at some standard distance from any view point and this will give us the shape of the visual image from that view point. So if we can infer the shape of an object from a set of visual data, it should also be possible to infer from the same evidence all the conditional visual experiences that are relevant to the truth of the statement that there is an object of a certain shape at a certain place.[4]

It seems that in principle there should be rules linking visual data to conditional predictions of visual experience and that we can dispense with beliefs about objective entities. However this is too fast; there is a crucial difficulty. Each predicted visual experience must be correlated with a possible view point, so that the different visual experiences are integrated via the spatial relationships of the different view points. But these positions can only be specified relative to a three-dimensional coordinate system which must be realized in some concrete structure. So what gave promise of providing rules which avoid concepts of objective entities fails. Objective concepts are implicit in the conceptual apparatus for using two-dimensional geometry to deal with three-dimensional objects. In addition, the foregoing treatment of visual experience was extremely schematic, passing over, for instance, stereoscopic phenomena and various other cues of depth. Also by dealing only with visual experience we not only simplified the problem but also selected the kind of experience that can most easily be subject to rules. When we take this into account and allow that experiences of other kinds, such as tactile and kinaesthetic, are relevant to the truth of a belief about a material object, we recognize that there are many further barriers to arriving at experiential expectations without employing beliefs about objective entities.

Our ability to think about the concrete world has undoubtedly been enhanced and systematized by the development of geometry and the techniques it gives us. A plan, elevation and section is sufficient to precisely specify some regularly shaped objects, provided the nature of the regularity is specified or understood. Indeed, a sufficient number of parallel sections can represent, to any desired degree of accuracy, any object of any shape. These techniques again suggest that our three-dimensional concepts of the shapes of material objects might be reduced to two-dimensional shapes which can be directly represented in visual imagery. But this account again passes over the crucial point. A set of sections only specifies a three-dimensional shape when we understand the relationships of the sections to each other and while this may be done by another appropriately oriented section, we still need to understand the notion of spatial orientation. Three-dimensional structures cannot be reduced to two-dimensional ones, even at the visual level. Ultimately we come back to our conception of material objects. Having these concepts certainly involves our ability to recognize conditional experiential expectations as implicit in our objective beliefs; but no set of these expectations can be identified with an objective belief. That we have the mental capacity to make these connections must undoubtedly be explained via the evolution of human neural systems. However the distinctively philosophical issue is to say why we ought to use the conceptual system that we find comes naturally to us.

3.4. *Our Rationale for Trusting Perception*

To predict we need to detect patterns in what is given in experience. We have seen that it is not possible to do this by dealing with experience directly, whereas it is if we take experience as originating from an objective world. The most direct way of arriving at beliefs about such a world is by trusting what comes naturally in perception, but we cannot ignore the possibility of some other way of systematizing experience. Perhaps experience is indeed objectively caused but in some way quite different to the way we take perception to work. This, of course, is to some extent what we do believe; for we allow that some perceptually based beliefs need to be rejected or modified. However in doing so we still put *prima facie* trust in perception and believe that it usually leads to true belief. What we do reject is that perceptual beliefs are caused by an objective world which is generally and radically different to the way we in fact take it to be and that this is due to some mechanism that is hidden from us. But once we grant the possibility of such deception, we must ask why it is reasonable to trust perception, rather than allow that the objective patterns may be quite different. The answer is this. To have an actual alternative method of arriving at objective beliefs from experience we would need a coherent body of rules which either rest on, or imply, an account of how the hidden mechanism work. But because there is no limit to the numbers of such possible mechanism and because each is, by hypothesis, hidden we could only guess at the nature of the mechanism that operates.

That our experience is naturally integrated by beliefs about a three-dimensional objective world is central to the rationale for our *prima facie* trust in these beliefs. This is not to say that we ought simply to believe what comes naturally. (Indeed a good deal of work in psychology shows there are a variety of irrational patterns of though that come very naturally to all of us.) As with induction, what makes our *prima facie* trust of perception rational is that it is a possible, viable way of trying to do what we want to do; while any other way of trying would negate the very point of intellectual endeavour. To seek anything means aiming to be an effective agent and this requires prediction. Rational predictions require the identification of patterns in the data. To try and identify such patterns without placing any trust in perception would once again involve trusting arbitrary guesses and so be irrational.[5]

4. OBJECTIVE PROPERTIES

The foregoing provides a context in which to return to the key issue: how to select between the multiplicity of hypotheses with differing predictive content that must always be compatible with any finite body of data. I am suggesting that we should focus on the nature of the objective patterns asserted by hypotheses, rather than on their linguistic expressions. A minimal requirement is that prediction be based on the projection of patterns that have persisted through known facts, but we need to clarify the standards by which we judge whether a pattern remains unchanged. As Goodman's work indicates, the notion of projection can only have significance given the objective properties and relations which we take to be exhibited in a pattern. We must allow that a property can be compounded out of others, but we need criteria for distinguishing which properties are more basic than others. Then we can go beyond (2a) in section 2.2 and formulate a viable requirement as to how predictive and evidential conjuncts (see (1a) in section 1) can be legitimately combined:

(4a) The two components of the overall pattern must each exhibit the same pattern with respect to those properties which we have reason to take as more basic than any others that might be employed in pattern identification.

This gives content to the notion of projection without change of pattern and induction can then be equated with this.

While some properties (like velocity) necessarily involve the change of other properties, the possibility of identifying a pattern of change presumes what might not have changed. Once it is possible to order property-concepts in a system which takes some as basic and the others as reducible to these, we can take change with respect to the basic properties as a non-relative standard of change (since the use of property-concepts which are reducible to others is no more than a convenience and has no ontological significance). We must predict changes in basic properties only if a pattern of such change has persisted throughout what has been observed. The whole point of induction is to avoid arbitrary guesses and this means expecting change only where it will occur if an extant pattern persists. If science establishes an ontologically significant system of properties, we should, in the last analysis, judge the persistence of a pattern relative to basic properties.

Let us consider our rationale for attributing properties to objects; for this may throw light on our rationale for taking some as more basic. I have argued for the thesis that our perceptual beliefs have a *prima facie* warrant and that we must take them as beliefs about an objective world. In adopting such beliefs we attribute properties to objects and the rationality of the beliefs implies the rationality of the attributions. This view of our knowledge of the world has some obvious implications that we should bear in mind. Firstly, our concepts of properties must be grounded in experience, since our beliefs which attribute properties are. On the other hand our most immediate beliefs are defeasible and involve interpretation. Consequently we cannot expect that the property-concepts which come most immediately from experience are thereby the most basic nor that they are irrevocable. What is given in experience may suggest a concept to us, but just as the beliefs involving that concept are subject to revision, so is the concept itself.

In considering psychological nominalism, I have argued that we are able to detect resemblances between the quality of experiences on different occasions. If we are to take experience as a guide to what is objectively the case, perceived similarity must count as a *prima facie* indication of objective similarity. Since the objective similarity of instances is just what constitutes a property, we must take perceived similarity as a *prima facie* ground for identifying a property. However the possibility of reinterpretation means that this does not settle the matter. The point of property-revision is to identify, more adequately, what the objective properties are. So it is the scientific, reductive systematization of properties, not epistemological priority, that we must take as ontologically significant. This is born out by the simple and obvious fact that different attributions of the same property are not always based on perceived resemblances and often on quite different sensory modalities. Let us, then, explore the different kinds of concepts we employ in thinking about the objective world.

4.1. *Concepts and Dispositions*

Since predictive content is a prime concern in adopting objective beliefs, we must consider the contribution that a concept makes to this enterprise in order to understand the grounds for employing or revising it. Once we start to reflect on the nature of perception and try to isolate what is actually given in experience, we may discover that modifying a concept that comes naturally may enhance the power of our system of predictive beliefs. The simplest revisions are a part of everyday thinking, but as the revisions become more radical and extensive the system that results will take on, more and more, the characteristics of theory. My thesis is that theorizing is continuous with the kind of revision that is involved in critical reflection on everyday beliefs.

While particular facts can falsify any generalization, any perceptually based belief about a singular matter of fact will be rejected if it conflicts too radically with other well established beliefs. However we try to make sense of such rejections. We may well challenge the memories involved; but there can be other sources of error. What memory primarily records is not experience itself, but an objective belief which it grounds; and if the belief is ill-grounded, accurate memory is no remedy. Questions about memory and interpretation are hard to untangle; but to deal with the latter we must try to get at what we have actually experienced on some occasion. This is difficult and must rest on circumstantial evidence. Our memory of objective beliefs formed, analogies with similar cases in which we may be more attentive to the experience itself and our best beliefs about the process by which perceptual beliefs arise from experience are all involved. It is clear that a variety of beliefs of different kinds are interdependent and that coherence considerations are important in the appraisal of even those beliefs closest to experience.

We revise not only singular beliefs but also whole categories of beliefs. Because these latter involve changes in the way we interpret experience, they are of especial interest. When beliefs that come naturally to us from experience are subject to critical appraisal, we must draw on our understanding of the processes of perception. This will often oblige us to seek an explicit formulation of notions which, until then, were latent and not well articulated. These beliefs, like any others, will then be open to critical appraisal and will sometimes be revised.

To understand the nature of such revisions, we must recognize how pervasive dispositional notions are in our understanding of all properties that are not spatio-temporal. It is not much of a simplification to take temporal order and the simultaneity of adjacent events, as given in experience. In this essay I will leave it at that. By contrast, our three-dimensional spatial concepts are not so simple; but they are obviously related to what is given in experience. Our beliefs about material objects provide expectations, even when the object remains unchanged, by enabling us to anticipate how experience will change as we move relative to it. I have argued that while our concepts of the shape of a concrete object is not reducible to visual images, it is nevertheless reasonable to take the structure of an image as isomorphic with an aspect of the shape of the concrete object that caused it. If this is correct, the whole system of visual experiential expectations that is part of our conception of the shape of an object is, in some sense, 'like' the object. However this cannot be said of non-spatiotemporal concepts and when it is recognized that our understanding of these properties is via our beliefs about dispositions, the way one property can be reduced to others becomes quite clear. The point is developed in Section 5.2 below.

4.2. *Colours and Vision*

We may be tempted to think that properties such as colours are not dispositional in the way that, for instance, malleability is. It may seem that colour perception is direct acquaintance with a property of an external object. However this cannot withstand critical appraisal. We must allow that the appearance of an object changes when it is illuminated by light that passes through a coloured screen. In these cases it is not easy to make sense of the thesis that the object itself has changed; however if we take the experience of colour to be due most immediately to the character of the light which impinges on the eye, it is easy enough to subsume the changed appearance under a regularity. On this account describing a surface as having a certain colour is attributing a dispositional property to it.

Our general understanding of visual perception, which may always be relevant when it is necessary to critically appraise a perceptual belief based on sight, has emerged over centuries and involves a substantial revision of the ideas that initially seemed most plausible. Most classical Greek thinkers thought of rays of sight extending from the eye to the seen object. This point of view was enough for the development of geometrical optics and theory of perspective from the time of Euclid to the Medieval periods; for the geometry was essentially the same whether the straight lines represented rays of sight from the eye or rays of light to the eye. The notion that something goes forth from the eye when we look at something is very natural. However there are problems. For instance, how does light come into the story? According to Aristotle, who accepted the rays of sight theory, transparent objects are, of their nature, only potentially transparent and it was the activity of light which actualized this potentiality. The transparency of water, air etc. consisted in the ability of our rays of sight to pass through them to an object which is thus seen (McKeon,1941, 418b04 *et seq.*). This is in accord with the most obviously relevant facts, but questions remain. We are told little about the nature of light and nothing about how its presence facilitates sight through certain substances. Nor are we told, by Aristotle or anyone else, how it is that information is able to travel back up the rays of sight to us. Obviously it is the character of the surface and its shape that causes our perceptual beliefs, so the question is how the rays of sight, going from us, enables the object out there to affect what happens in us. These problems are avoided when we take visual experience to be caused by rays of light entering the eye. This thesis had been entertained by a few Greek and Medieval thinkers (see Crombie,1959, vol. 1 pp.99-113) but was not widely accepted until it had been expounded by Descartes. There is perhaps the appearance of a problem in explaining how light rays conveniently travel directly to the eye of the observer; but this vanishes once it is allowed that every point of any illuminated surface either emits or reflects light in every direction. On this story we are left with just light and its effects. There are no rays of sight and a transparent body is simply one through which light can travel.

On our current understanding of vision it is easy to accept that the colour of a surface is a dispositional property, one of the factors that determines the character of the light that enters our eyes, and so, in turn, influences the quality of our visual experience. While colour words can be used in describing the quality of this experience, there is no sense in which it is a copy of the property of the object. The conception of colour is derived from experience, initially uncritically. However with time the naïve notion that something from us reaches out and detects a quality of the object was seen to be inadequate. A rather different account of visual experience also accommodated the most obvious features of seeing that something is, say, red, but avoids inconsistencies and gaps in the earlier view. Nevertheless perceived resemblance, in the absence of other, overriding considerations, remains an indication of sameness of an objective property. Consequently the greenness of an object (at a certain time) is a more basic property than its grueness. Granted the principle (4a), that the patterns which we inductively project must be the same pattern with respect to the most basic properties, we have here a solution to Goodman's riddle. This has emerged from an examination of the way property-concepts are based on experience; for this in turn throws light on concept revision and the ordering of properties with respect to how basic they are. In the next section I will argue that this systematization plays a crucial role in the development of theory and in the rationale of theory acceptance.

5. INFERENCE TO THEORY

It is widely held that theory is accepted because of its explanatory power and that the criteria that establish explanatory power are quite distinct from those that guide induction. This implies that there are two separate modes of ampliative inference. This duality is sometimes avoided by taking inference to the best explanation to be basic and induction to be subsumed under it. David Armstrong takes this position and sees the justification of induction as deriving from that of inferences to the best explanation. However, he has little to say about the rationality of these latter inferences; claiming only that it is analytic that they are so:

> To infer to the best explanation is part of what it is to be rational. If that is not rational, what is?' (Armstrong 1983, p.59).

This, even although he dismisses as unsatisfactory a similar argument by Strawson for the rationality of induction.[6]

I suggest that we go in the opposite direction to Armstrong and some others, and take inferences to the best explanation to be justified because they are included within the inductive method. It is not difficult to argue that theorizing both unifies and extends the scope of our system of predictive belief. Thus it is possible to argue that it is only by theorizing that we can detect the most extensive and unified pattern in the facts. If the account that I gave above of induction is correct, it follows that theorizing is a necessary part of inductive reasoning.

This approach provides a defence of scientific realism against the arguments of Bas van Fraassen (1980). He denies that there is any rational obligation to believe the best explanation and holds that science is concerned only with the empirical adequacy of theory and not with its truth. An important part of his argument is the claim that the empirical content of a theory can be clearly demarcated from the rest and can be asserted separately. If he is correct in this, exactly the same empirical predictions result from something less than the assertion of the theory. It would then seem reasonable to see the explanatory role of theory as only contributing to a human psychological need: our sense of having understanding. It is certainly an empirical fact that humans have such a need, but it is equally a fact that what provides intellectual satisfaction to one does not to another and may well conflict with the scientific consensus. If having some degree of confidence in the truth of theories is part of scientific method, then it is essential to show that theory does something more substantial than meet a psychological need. This is highlighted by van Fraassen.

I hold that the claim that a theory's empirical content can stand independently of accepting that the theory is at least approximately correct, is mistaken. I have argued elsewhere (Clendinnen 1989), that our hypotheses about what we can not directly observe contribute very substantially to our ability to predict. What I want to do here is show that theorizing is continuous with the kind of thinking that we must employ in our everyday dealings with the world and in particular with the way we critically appraise our perceptual beliefs.

5.1. *Hidden Structures and the Integration of Belief*

Since the earliest days of natural philosophy, hypotheses about unobservable entities have been proposed in order to make sense of phenomena that were otherwise puzzling. Lucretius summarizes some of the most telling examples of such speculations by atomists from the fifth century B.C. on. We can feel the wind and its effects, smell a crushed herb, note that clothes dry or grow moist in different conditions, observe that a wedding ring grows thin over the years and that other objects wear away (Lucretius 1951, pp.35-36). All these can readily be subsumed under familiar regularities if we allow for the movement of particles which are too small to see. Note that we need no speculations about unobservables to ensure the phenomena in question are subsumed under general patterns that facilitate prediction. An induction on what we observe enables us to state a generalization about the changes we will see in wedding rings worn continuously over the years; but if we leave it at that we are have only a generalization of very limited scope. The policy it exemplifies will require the continual use of clauses stating *ad hoc* exceptions. Metallic objects usually retain their shape and dimensions unchanged, but sometimes not. The list of deviant contexts includes some in which observable causes act; like the use of a file or the application of heat and hammer blows. But it also includes the case of regularly worn rings.

We aim not simply at subsumption under regularities but also at subsumption under regularities that are as extensive as possible. If Hempel's covering-law account of explanation is correct then, as he allows, we must seek explanations of less general laws in terms of more general ones; as well as the explanation of singular facts. That is why the atomists' explanation offers a deeper explanation than just saying what we can generally observe of wedding rings and such like. However what micro-theories offered was limited so long as the properties attributed to microentities were simply those which were suggested directly by experience. Indeed the uncritical acceptance of such properties may preclude micro-hypotheses altogether. If we take the solidity of objects in a simplistic way we may be led to assume that they are solid through and through, so concluding that they are homogeneous and lack all structure. Even when atomists avoided this, and did speculate about microparticles, they assumed only those properties which we naturally attribute to the larger objects we perceive.

Our naïvely empirical conception of objective properties are by no means sacrosanct. What we perceive of a solid object in no way requires it to be homogeneous at the microlevel. The notion that it is, is no more that an easy, although illegitimate, extrapolation. What we can observe is that the bodies we call solid resist penetration and distortion when in contact with similar bodies and other such dispositions. The legitimate conception of solidity need go no further than this. Having accepted this, it is a small step to recognize that we are by no means committed to the applicability of perceptually based properties to microparticles. Our decisions as to what properties to posit should be guided by what they contribute to the overall systematization of predictive belief.

5.2. *Theorizing and Property Revision*

Theorizing about microstructures only resulted in a dramatic increase in our predictive power when it was combined with more sophisticated conceptions of basic physical properties. The Newtonian concepts of mass and force were crucial. The development of concepts of electrical charge and magnetic force and later the concepts of quantum theory, resulted in microhypotheses providing not only impressive unification of laws but also predictions which would otherwise never have been made.

These developments contributed to both of two desiderata that can pull in opposite directions: that our system of predictive beliefs be well supported and also powerful. Theorizing, by reducing the set of laws which covers a certain range of fact, results in each law having a larger number of instances and so being better confirmed. At the same time, since the laws so identified are presumably more basic, there is hope that a new theory will cover new facts and facilitate new predictions. This plausible hope made it reasonable to be interested in theory before it actually facilitated innovative prediction. What has actually happened in science during the last couple of centuries immensely strengthens this rationale. Time and again a micro-hypothesis, put forward to deal with a few facts which were awkwardly outside the current system of laws, was found to lead to new, subsequently confirmed, predictions. So there is a second order induction of the kind to which Laudan pointed: that theorizing contributes to detecting the most comprehensive pattern in the data. Nevertheless seeking the most comprehensive pattern is constitutive of induction. So an argument that establishes the rationality of induction thereby establishes the rationality of theorizing.

Theorizing, therefore, involves the development of a system of sophisticated properties. However there is a question as to just what we mean in holding that one property should be replaced by another. In the case of spatio-temporal properties it is easy enough to see how an hypothesis about the properties of a microstructure can determine the properties of the object constituted by that structure. Something being composed of parts with certain shapes and spatial arrangements clearly entails its shape and position and its motion and change of shape is determined by the motions of those parts. Where properties are dispositional, it is again not hard to see how property-reduction works. If X having the property P is constituted by X being disposed to certain changes in certain contexts, having disposition D, say, then having property P will be entailed by having property Q if this also results in disposition D. Put thus, it sounds as if the procedure is trivial and pointless. However it certainly is not if property Q is a structural property; that is X having Q consists in it having certain kinds of parts related in certain ways. The hypothesis that Q is the same property as P will be supported if the laws which spelled out D may be derivable just from the hypothesis and laws that deal with the kinds of entities that it asserts are X's parts. The effect of this hypothesis will be to show that laws invoking property P are unnecessary. The hypothesis that X has the property Q is in part a claim that parts are arranged in a certain spatial structure, but it is also a claim that the parts are of certain kinds. The point I want to stress is that the development of the property-concepts that are essential to effective microtheorizing must both build on the properties which arise naïvely from experience and also involve a critical appraisal and modification of these concepts.

The concepts we have of colour are examples of those which come very directly from experience. There are two related lessons that can be learned by examining them. Firstly, that they are dispositional concepts; secondly, the concepts that we nowadays accept as unproblematic in fact rest on beliefs about what we can by no means observe. Since it is impossible to hold that we are directly aware of an objective quality in colour perception, we are pushed towards accepting the colours of objects as dispositional properties. This conclusion is made secure by our understanding that visual experience is caused in us by the light from objects that passes into our eyes. That this is so is almost a truism for the members of many present day societies, even for many who are generally innocent of scientific education. However a moment's reflection makes it clear that we can by no means observe the transmission of light and, as we have seen, for many centuries serious, thoughtful people accepted an account quite different to ours. More natural and naïve ideas about visual perception gave way to the modern view only as a result of strenuous searching for a thoroughly coherent and comprehensive account. Such an account was only to be had via speculations about what could not be directly observed. However this was nothing new, for the rays of sight postulated by the Greeks were no more observable than are rays of light. The crucial point is that our understanding of the mechanism of perception must go beyond what is open to observation; yet this understanding must play a crucial role in the critical appraisal of all our observation-based belief and in the modification of our property-concepts.

The properties that are employed in scientific theory are even more obviously dispositional than more naïve properties. There is a good case for holding that our understanding of a theoretical property is in terms of the laws which relate that property to others. To hold that we understand all objective properties thus would lead to an epistemic circle; however we have noted that some objective properties are understood via their relationship to the content of experience. Although an intrinsic property must be understood in terms of dispositions to produce other properties, there are some properties which involve dispositions to produce experience of a certain character. Likewise the spatial properties we attribute to objects are related to (although not reducible to) the structure of experience, in particular the two-dimensional structure of visual experience. It is thus possible for us to make sense of the attribution of objective properties via our expected and imaged experience.

Our beliefs about the objective world are based on experience; but inevitably involve integration and interpretation. Every belief about the objective world, however directly based on experience, is revisable; and the revision will be effected by considering how other experience bears on the belief and how the variety of our experience is to be best integrated. This is true at every level and the distinction between everyday belief and theoretical belief is but a matter of degree.

5.3. *Objective Belief and Experience*

The perspective that I propose agrees with van Fraassen's in taking the predictive capacity to be the prime concern of theory, however I see the realistic interpretation of many theories as the only effective way of integrating phenomena under regularities. Suppose we could identify all the statements derivable from a theory which were observationally verifiable. This set of sentences would then be the empirical content of the theory, and could be organized systematically. Craig's theorem assures us that it is axiomatizable, albeit in a very complex system. However it is generally agreed that the only practicable way of formulating the Craigian axioms is by starting with the original theory. What discussion has centred on is whether this is of ontological significance.[7] I hold that the integrative contribution that any belief makes to our system of beliefs is relevant to our evaluation of its truth. The continuity that this implies between everyday and theoretical propositions is obscured if observational reports are taken as unproblematic. It is interesting that van Fraassen does just this. He insists that his line of argument could not challenge realism with respect to everyday entities, saying that the suggestion that it could presupposes the decisively discredited sense data theory (van Fraassen 1980, p.72). However denying tenets of sense data theory — that beliefs about objective entities are typically inferred from beliefs about experience and that these latter are incorrigible—does not avoid the fallibility of our most basic empirical beliefs. By taking perception for granted van Fraassen ignores the possibility that the criteria used in its appraisal may have a wider application. Further, the assumption that observation gives us direct knowledge sustains the notion that the integrative role of a belief is purely pragmatic and not relevant to truth.

The perspective that I have tried to develop in this essay allows that we have an innate capacity and tendency to interpret our experience as originating from an objective world. However this inclination cannot be enough to ground our confidence that there is such a world, or that it has the character that we take it to have. Whatever is initially taken for granted deserves to be appraised and re-evaluated. This is the task of philosophy. So from a philosophical perspective we are obliged to critically examine the belief strategies that come naturally. I have argued that when we undertake this appraisal, we can recognize the rationality of taking our innate inclinations to objective beliefs as a starting point, albeit a fallible one. It is not that there is, or could be, any guarantee that this policy will usually result in true belief; but we can compare this policy to others. When we do, we can see that all possible alternatives are irrational in a way that the empirical approach is not. So the policy of taking our experience as arising in a systematic way from an objective world offers a way, and the only rational way, of seeking beliefs that will be successful.

What is of crucial concern to us is that our beliefs provide a good guide to practice and I have argued that it is relative to this goal that the empirical method can be shown to be a rational policy. However we would also like to be assured that we have representations, the best of which correspond well with reality. Here my argument that our beliefs must take the form of representations is relevant but does not settle the matter. It clearly possible that a representation that fails to correspond to anything objective may nevertheless guide practice so that it usually succeeds. On the other hand it is clear that being guided by a representation that is largely correct will lead to considerable practical success. Given that it does not even make sense to suppose that we might have direct access to reality, taking a representation that works well as probably fairly accurate is the only reasonable way of forming a view of the nature of reality.

IN CONCLUSION

To sum up; the naïve empirical beliefs that must be our starting point are revisable and revisions should be guided by what will contribute most to the integration of experience. We want to identify patterns which cover the widest range of experience. I have argued that it is reasonable to try to achieve this by seeking a set of laws each of which accounts for as much of experience as possible. In the first place, some interpretation of experience is essential if we are to recognize any kind of patterns; and having taken this step it is plausible that the wider the generalizations we employ to account for what has been observed the more basic they will be and so likely to cover more of the as yet unobserved. This conjecture has been strongly confirmed by the achievements of science. There is a bonus with this policy: the wider the scope of each law, the more confirmatory instances and the better it is grounded. So our intuitive tendency to 'make sense' of more or less isolated phenomena by seeking to subsume them under generalizations which are at least reasonably wide, is vindicated by being required by our rational, inductive method.

Coherence considerations are involved in the appraisal of all beliefs. They all involve interpretation and we are always concerned to seek the system which, while evidentially well supported, is as predictively powerful as possible. So we prefer interpretations that maximize these desiderata. We do this in appraising individual beliefs, in critically modifying our property-concepts, in adopting beliefs about parts of reality which are more remote from experience and in seeking to understand the mechanism of our own perceptions. As we have seen, these are all interrelated. While we can draw a rough and ready distinction between what we ordinarily do and do not call theoretical beliefs, no precise line is possible. In particular, if we insist that theory starts when beliefs involve interpretation and go beyond what is given in experience, then we are obliged to say that all beliefs are theoretical.

University of Melbourne, Victoria

NOTES

* I would like to express my gratitude to the anonymous referee for this volume whose comments led me to revise the paper considerably.

[1] This occurs in his response to a paper by S. F. Barker, and Peter Achinstein (1968) in which they argue that 'grue' is positional. The core of their argument being that in a painting a grue object could not be represented naturally. The best that could be done would be two paintings, each including some indication of the date. However Goodman insists that there are no natural representations. In particular, we can not hold that a natural representation of the colour of an object is indiscriminable from it; since green paint will exactly match very few green objects (1968, p.163). Consequently he holds that there is no reason why grue objects might not, by convention, be represented by, say, a hatched shading. He here inserts a footnote including the above cited passage.

[2] See Clendinnen (1977).

[3] While I see Haack's work as immensely valuable I remain uneasy with her account of our rationale for granting a *prima facie* warrant to perceptual beliefs. What follows is an attempt to provide an alternative.

[4] The step from visual data to three-dimensional shape could not be according to an algorithm, even assuming that the former can be adequately described. However granted the capacity to recognize which experiences are compatible with an objective belief, it will be possible to accumulate supporting experiential evidence for the latter. Only in the rare cases where there is some contrary evidence to a perceptual belief will we need to appraise it and when we do, we will look for an objective belief which will predict the experiential data. The fallibility of this step for the realist implies no obstacle to the phenomenalist, beyond requiring that s/he allow that the inference from the data of experience to prediction is fallible.

[5] As I have indicated, I believe that this is very much the justifying argument that Sellars intended in his (1979). He takes effective agency to be the goal of the empirical method, by which he means what is specified by the principle of induction and three principles which assert the rationality of trusting perception as well as the memories and introspections that are implicit in it. He says that being an effective agent can be shown to require the empirical

method as means (§§68, 82). If this is read to mean that making predictions that result in successful action requires the use of the empirical method, it is incorrect. However I assume that by 'effective agency' Sellars meant to require that the action was rational as well as successful. If so, he is claiming precisely the justifying relationship for which I have argued. He points out how the empirical method does systematically provide cognitive maps to guide action. He does not actually argue that any other way of trying would be unreasonable, but I suspect that he meant to take this as read. I have argued at more length for my interpretation of Sellars' argument in 'Sellars and the Grounding of Empirical Belief', available as the *Victorian Centre for the History and Philosophy of Science Preprint* **3/92**.

[6] That such justificatory arguments are unsound is clear as soon as we inquire whether 'X is rational' entails 'X ought to be done'. If it does, the claim about the meaning of 'rational' is incorrect; whereas if there is no such entailment, we are left with the question 'Why do what is rational?'.

[7] See Smart 1963, p.32 and Hooker 1968, pp.67-68.

REFERENCES

Aristotle (see McKeon, 1941).

Armstrong, D. M. (1983) *What is a Law of Nature?* Cambridge, C.U.P.

Clendinnen, F.J. (1977) 'Inference, Practice and Theory', *Synthese* **34**, 89-132.

Clendinnen, F. J. (1982) 'Rational Expectation and Simplicity' in *What? Where? When? Why? Essays on Induction, Space and Time, Explanation.* McLaughlin, R. (ed.) Dordrecht, Reidel, 1-25.

Clendinnen, F. J. (1986) 'Induction, Indifference and Guessing', *Australasian Journal of Philosophy* **64**, 340-344.

Clendinnen, F. J. (1989) 'Realism and the Underdetermination of Theory', *Synthese* **81**, 63-90.

Crombie, A. C.(1959) *Medieval and Early Modern Science* , vols 1 & 2, New York, Doubleday Anchor.

Barker, S.F. and Achinstein, P. (1968) 'On the New Riddle of Induction', in Nidditch (1968) 149-161. (First published *Philosophical Review* **69** (1960)).

Goodman, N. (1965) *Fact, Fiction, and Forecast*, Indianapolis Bobbs-Merrill. (First edition, 1955).

Goodman, N. (1968) 'Positionality and Pictures',in Nidditch (1968) 162-64. (First published *Philosophical Review* **69** (1960)).

Goodman, Nelson (1961) 'Safety, Strength, Simplicity', *Philosophy of Science* **28**, 150-51.

Haack, Susan (1993) *Evidence and Inquiry: Towards Reconstruction in Epistemology*, Oxford, Blackwell.

Hooker, C. A. (1968) 'Five arguments Against Craigian Transcriptionism', *Australasian Journal of Philosophy* **46**, 263-276.

Laudan, Larry (1987) 'Progress or Rationality? The Prospects for Normative Naturalism', *American Philosophical Quarterly* **24**, 19-3.

Lucretius (1951) *The Nature of the Universe*, R.E. Latham (Trans.), Middlesex, Penguin.

McKeon, R.(ed.) (1941) *The Basic Works of Aristotle*, New York, Random House. (References are to the Bekker edition pagination, retained by McKeon.)

Nidditch, P.H. (1968) *The Philosophy of Science*, Oxford, O.U.P.

Salmon, W.C. (1963) 'On Vindicating Induction' in Kyburg, H. and Nagel, E. *Induction: Some Current Issues*, Middletown, Conn., Wesleyan University Press. pp.27-41.

Salmon, W. C. (1966) 'The Foundations of Scientific Inference' in Colodny, R.G. (ed.) *Mind and Cosmos* Pittsburgh, University of Pittsburgh Press.

Sellars, Wilfrid (1956) 'Empiricism and the Philosophy of Mind' in *Minnesota Studies in the Philosophy of Science* , vol. 1; *The Foundations of Science and the Concepts of Psychology and Psychoanalysis*, Feigl, H. and Scriven, M. (eds.) Minneapolis, University of Minnesota Press. pp.253-329.

Sellars, Wilfrid (1973) 'Givenness and Explanatory Coherence', *Journal of Philosophy* **70,** 613-624.

Sellars, Wilfrid (1979) 'More on Givenness and Explanatory Coherence' in *Justification and Knowledge,* Pappas G. S. (ed.) Dordrecht, Reidel. pp.169-82.

Smart, J. J. C. (1963) *Philosophy and Scientific Realism,* London, Routledge and Kegan Paul.

Triplett, Timm (1990) 'Recent work on Foundationalism', *American Philosophical Quarterly* **27**, 93-116.

Van Fraassen, Bas C. (1980) *The Scientific Image*, Oxford, Clarendon Press.

LEN O'NEILL

INDIFFERENCE AND INDUCTION

Many find that there are at least some situations in which the notorious Principle of Indifference provides a reasonable basis for a distribution of probabilities, construed as degrees of confidence. For example, knowing only that there is a bean in one of three boxes, but ignorant of any information as to which box it is in, the only reasonable distribution of degrees of confidence, is a uniform one of 1/3 to each possibility. It is the purpose of this paper to consider what might be offered by way of a rationale for the principle and how two objections to the principle might be handled.

The rationale offered, in Section One, operates, like the Dutch Book argument, in terms of betting odds rather than degrees of confidence. It is decision-theoretic in nature, specifically a minimax loss argument. Since the context in which the Principle of Indifference is one, in decision-theoretic terms, 'under uncertainty' (or ignorance), a minimax argument is specially appropriate. It is conceded that such arguments are not apodeictic, but, nonetheless, are widely held to constitute one reasonable determinant of choice under uncertainty. Section Two examines the objection from epistemological magic, which holds that it is impossible to 'conjure' reasonable degrees of belief out of the 'thin air' of ignorance. This objection is dealt with by arguing that there is a sort of consilience between it and the rationale offered in Section One. That rationale, strictly speaking, could only justify betting odds, and not degrees of confidence. This grounds a separation of betting odds from degrees of belief and thereby allows the Principle of Indifference to be consistent with the objection from epistemological magic by restricting itself to the distribution of betting odds.

In Section Three, the more complex objection from non-linearity is examined. This objection holds that logically equivalent formulations of the same information in terms of different units yield inconsistent recommendations according to the Principle of Indifference. The strategy in this section is not to rebut the objection directly but to link it to the problem Goodman's 'grue' predicate raises for the Principle of Induction. A recent attempt by George Schlesinger[1] to tie the Principle of Indifference more closely to inductive inference is also examined in this section.

I

In dealing specifically with the classical formulation of the Principle of Indifference in terms 'equal possibilities', Wesley Salmon notes:

P.J. Riggs (ed.), Natural Kinds, Laws of Nature and Scientific Methodology, 93–102.

> Another objection rejects any rule that pretends to transform ignorance into knowledge ... According to the principle of indifference, this kind of knowledge [of probabilities] can result immediately from our ignorance of reasons to regard one occurrence to be more probable than another. This is epistemological magic.[2]

More recently, Michael Resnik has made a similar point:

> If there is no reason for assigning one set of probabilities rather than another, there is no justification for assuming the states are equiprobable either ... If every probability assignment is groundless, the only rational alternative is to assign none at all.[3]

Using 'conditions of ignorance' to denominate the conditions in which the Principle of Indifference is intended to apply, the following captures the conflict between (A) the epistemic version of the Principle of Indifference (i.e., in terms of distribution of confidence) and, (B) the objection from epistemological magic:

(A) In conditions of ignorance, it is rational to distribute one's degrees of confidence equally over the alternative possibilities.

(B) In conditions of ignorance, there is no rational distribution of degrees of confidence over the alternative possibilities.

The familiar Dutch Book argument might be described as a kind of decision-theoretic defence in conditions of ignorance of having one's degrees of confidence conform to the axioms of the probability calculus. The argument purports to show that failure to so conform generates an incoherent belief set in the sense that were one to accept all bets at odds reflecting these degrees of confidence, bets could be so placed with you that you must sustain an overall loss. Conformity to the calculus is argued for in terms of possible losses of what you value. One might even see it as a minimax loss argument, that is, as an argument which puts an upper limit on how much you can lose for a fixed total stake as a result of an opponent's mathematical ingenuity. The more your degrees of belief diverge from the axioms, the more you can be forced to lose. Coherence minimizes this by getting it down to zero. It serves as a constraint on degrees of belief even under conditions of ignorance. Similarly, there is a version of Pascal's wager that operates under conditions of ignorance. The loss consequential upon a mistaken belief in God is relatively slight, whereas a mistaken disbelief would be a great loss indeed. By believing in God you have much to gain and little to lose, whereas disbelief offers the possibility of little gain and large loss.

Consider now the Principle of Indifference. In conditions of ignorance, the principle proposes a far tighter constraint on one's system of belief than mere coherence. The latter requires that your probabilities sum to one and hence, in our three bean example, a distribution of 1/2, 1/8, and 3/8 is acceptable. The Principle of Indifference would disallow any but 1/3, 1/3 and 1/3. What decision-theoretic advantage could this have?

Coherence, one might say, protects you from the mathematical astuteness of an opponent. It does nothing, of course, to protect you against advantages of empirical information your opponent might have. In conditions of ignorance you are particularly vulnerable to such an opponent. Suppose the opponent *knows* in which box the bean is. You could, say, offer zero odds that it is under box X and retain coherence by appropriately lengthening the odds you offer on Y and Z. If, indeed, the bean is under X, you thereby stymie the opponent, but if the bean is not under X, you face a heavier loss. There is no way, in conditions of ignorance, you can guarantee to have no loss, the most you can hope for is to keep it to a minimum. Granted you keep your acceptable odds coherent, there seems to be only one way to further secure a minimization of your possible loss. If you offer the *same* odds, of two-to-one, on each of X, Y and Z, then the maximum the opponent can win is twice his or her stake. Any divergence from this opens up the *possibility* of greater loss.

It is appropriate to appeal to a minimax loss strategy in support of the Indifference Principle. The context in which that principle operates, conditions of ignorance, corresponds to decision 'under uncertainty', where minimax strategies are appropriate (as opposed to 'decision under risk', where probabilities are available).

Minimax loss arguments are not apodeictic and are subject to various problems. Such arguments are essentially conservative and, at least in some contexts, may not be uniquely rational. However they have been given sufficient philosophic prominence, most notably by John Rawls,[4] to be used in this justificatory role.

II

This minimax defence of the Principle of Indifference appears to make it vulnerable to the objection from epistemological magic, for it argues that degrees of confidence can be justified in conditions of ignorance. It recommends accepting a certain distribution of betting odds and this, it is reasonably held, constitutes an acceptance of a parallel distribution of degrees of confidence. Hence, there is conjuring of favoured epistemic attitudes out of the thin epistemological air of ignorance. To deal with the objection we need to unpack (A) into two parts (A_1) the 'behavioural principle of indifference' and (A_2) 'the personalist thesis'. This yields the inconsistent triad A_1, A_2, and B:

(A_1) In conditions of ignorance, it is rational to distribute one's betting odds equally over the alternative possibilities.

(A_2) Rationally acceptable betting odds measure degrees of confidence.

(B) In conditions of ignorance, there is no rational distribution of degrees of confidence over the alternative possibilities.

If (A_2) is accepted then the 'behavioural' principle of indifference (A_1) becomes equivalent to the Principle of Indifference in terms of degrees of confidence (A). The inconsistency between (A_1), (A_2), and (B) can be resolved by rejecting any of the three components. The minimax justification offered in the preceding section strongly supports a rejection of (A_2). Whether or not there is rational belief in conditions of uncertainty, it is widely held that there is rational action in such conditions. Dominance arguments in Prisoners' dilemmas, and elsewhere, and Rawls' use of maximin in the defence of his two principles have made decision under uncertainty widely known. On the other hand, rational action, it is plausible to hold, can provide access to an agent's beliefs and the strength with which he or she holds them. Decision under risk is admirably suited to this purpose, for in this situation the agent attends to his or her values and beliefs in determining his or her actions. But it is equally clear that rational action originating in conditions of uncertainty is ill-suited to providing access to a person's beliefs and the degree to which he or she holds them. Since under uncertainty there *are* no beliefs to play any role in determining action. This role is played by the preference ranking of outcomes alone, the chosen action can at most legitimate an inference to the agent's values and the relations between them and not at all to his or her beliefs. There is, of course, the important matter of how we are to determine whether the agent is acting under risk or under uncertainty, but the argument of this essay is not at any special disadvantage in this regard.

Our concern is with the special case where the action taken under uncertainty is that of accepting a bet, i.e., precisely where, by (A_2), some hope to gain access to degrees of confidence. But it is widely admitted that the way acceptable odds measure degrees of belief is not straightforward.[5] Other factors may be reflected in acceptable odds, such as an aversion to risk in general or to risk on the particular issue bet upon. Neither these considerations, nor our problem about betting odds accepted under uncertainty show the personalist claim to be unfounded. The personalist needs to abandon the quite general formulation of (A_2) and find appropriate restrictions for its operation. According to the foregoing argument one such constraint would preclude inferring an agent's degrees of confidence from the bets he or she will accept in conditions of uncertainty.

The strategy in this section has been to resolve the conflict between the Principle of Indifference (A) and the objection from epistemological magic (B) by drawing out a corollary of the maximin justification. Since this justification operates under uncertainty it cannot warrant an inference back to the agent's epistemic states. Hence it forces the rejection of (A_2). In so doing, it also forces us to distinguish the behavioural (distribution of betting odds) from the epistemic (distributions of confidence) principle of indifference. The latter, but not the former, is subject to the objection from epistemological magic. The former, but not the latter, is justified by the maximin argument of Section One. Hence, it is reasonable to bet at odds of 2/1 with respect to each of the cups, X, Y and Z, that it covers the bean. It is unreasonable to have degrees of belief about where the bean is. Insofar as the Principle of Indifference has an intuitive appeal, provision of a rational basis for it is limited to the foregoing. It counts in favour of the minimax justification that it provides the resources *both* to deal with the objection from epistemological magic and to provide a rationale for the Principle of Indifference.

III

The objection from non-linearity has been noted by many critics of the Principle of Indifference. Salmon provides a clear example:

> Suppose we know that a car has taken between one and two minutes to traverse one mile, and we know nothing further about the time taken. Applying the principle of indifference, we conclude that the probability that the time was between one and one-and-one-half minutes equals the probability that it was between one-and-one-half and two minutes. Our data can, however, be expressed in another way. We know that the average speed for the trip was between sixty and thirty miles per hour, but we know nothing further about the average speed. Applying the Principle of Indifference again, we conclude that the probability average speed was between forty-five and thirty miles per hour. Unfortunately, we have just contradicted the first result, because the time of one-and-one-half minutes corresponds to forty, not forty-five miles per hour.[6]

While there may be a need to tidy some details of Salmon's argument, the general problem is clear. It arises whenever we can express the same information in terms of units of different magnitudes which are related in a non-linear manner. Any information about the size, say, of pizzas can be expressed as size of radius (say, in centimetres) or of area (say, in square centimetres). But these two units are not in a *linear* relation, e.g. doubling the radius is not the same as doubling the area. A parallel to Salmon's example could readily be generated in terms of units of length and units of area. In generating the objection it is assumed that a principle along the lines of 'treat units of a given magnitude in the same way' governs the operation of the Principle of Indifference. The problem arises because different units, non-linearly related, yield different outcomes on this assumption. The assumption will be retained, but not justified, in this essay.

The line of argument of this section is that if we construe the objection from non-linearity as the problem of to which magnitude to apply the Principle of Indifference, it displays at least a formal similarity to the problem of determining to which 'property' (green or 'grue') to apply the Principle of Induction. If the similarity can be shown to be sufficiently great, then, first, so far as the objection from non-linearity is concerned, the Principle of Indifference is no worse off than the Principle of Induction and, second, the ways in which philosophers have sought to deal with the 'grue' problem may equally apply to dealing with the non-linearity objection.

A strong formal similarity between the 'grue' problem and the non-linearity problem is evident. The 'grue' problem rests on the fact that the information that all emeralds observed before year 2000 have been green is logically equivalent to the information that all emeralds observed before year 2000 have been 'grue'. This parallels the claim in Salmon's example that the same data can be expressed in terms of time taken or in terms of average velocity. The application of the Principle of Induction to the rival formulations yields both that it is reasonable to believe all emeralds are green and that it is reasonable to believe all emeralds are grue and offers us no ground for deciding between these. The application of the Principle of Indifference to the rival formulations of the car-trip data yields different probability ascriptions to the same proposition (recommends different betting odds, on the behavioural version of the principle) and offers no ground for deciding between these.

There are at least two differences between the two problems. First, the grue problem is concerned with qualities (or properties) whereas the non-linearity problem is concerned with quantities (or magnitudes). Second, the grue problem is readily generalized so that, in effect, the Principle of Induction legitimates all generalizations consistent with the data. Thus, we can replace 'blue' in the definition of 'grue' by any other colour legitimizing, in effect, that any post-2000 emerald will be any colour whatsoever. It is not evident that the non-linearity objection can be so generalized that not merely does it not decide between two alternative probability ascriptions but it legitimates all probability-ascriptions consistent with the data.

Goodman's strategy with 'grue' is readily extended to magnitudes. In the way that 'grue' is a function of time and colour, so an exotic magnitude 'units of glength' could be defined in terms of time and length. To be two 'units of glength' is to be observed to be one centimetre long before 2000 A.D. or else to be 2 centimetres long. More generally, to be x units 'glength' is to be observed to be x centimetres long before 2000 A.D. or else to be 2x centimetres long. Nor, if we have access to Goodman-type predicates, is there any difficulty in generalizing the non-linearity problem. To generate a Goodman problem, we need a function f^1 which concurs with a familiar function f over the cases so far observed, but otherwise diverges. To generate a non-linearity problem we need a function of f^1 which is non-linear with respect to a given function f. Area is a non-linear function of radius of a circle, but there are of course indefinitely many non-linear functions of the radius r, such as $\sqrt{r}$, log r etc., most of which are of as little interest as grue. Nonetheless they generate new probability assignments on the basis of applying the Principle of Indifference. For any desired probability assignment a function, probably Goodmanesque, can be found to yield it.

While holding back from the more radical claim that the two problems are in essence the same, perhaps enough has been said to warrant considering whether responses to the grue-problem provide instruction relevant to the non-linearity problem. Both require us to explain why some magnitudes should be favoured over others in the application of the respective principles. The responses to the 'grue' problem have been diverse. Some have argued that some grading of properties on a continuum of *simplicity/complexity* is needed for the operation of the Principle of Induction. Others have argued that some properties are *more basic* than others and, that on such views, Goodman's contention that blue can be defined in terms of 'grue' absurdly seeks to define the more basic (or simple) in terms of the less basic (or less simple). One might for example favour the application of the Principle of Indifference, in Salmon's example, to time rather than average velocity, on the ground that time is the more basic magnitude in terms of which average velocity must be defined.

One of the most instructive responses to Goodman's problem starts from noting that the critical notion is that of sameness. Induction seeks to generalize observed similarities. But what is a 'similarity'? Why is 'green' a similarity between observed emeralds and 'grue' not a similarity? Why do we generalize over the magnitude length but not over the magnitude of 'glength'? The notion of a unit of magnitude is also a notion of sameness - of 'same quantity as'. Why do we favour the units we do favour? Wittgenstein wrote:

> How could I show them that -as I should say- you don't really buy more wood if you buy a pile covering a bigger area? - I should, for instance, take a pile which was small by their ideas and, by laying the logs around, change it into a 'big' one. This *might* convince them - but perhaps they would say: "Yes, now it's a *lot* of wood and costs more"- and that would be the end of the matter. -We should presumably say in this case: they simply do not mean the same by a "lot of wood" and "a little wood" as we do; and they have quite a different system of payment from us.[7]

Quine asked why the 'quality-space' that is natural to us should 'correspond' to that of the outside world. His answer is the evident significance to survival of the ways in which we generalize. If we had found grue-like generalizations salient, and had generalized accordingly, then our survival would have been dramatically threatened. The successful operation of the Principle of Induction presupposes an innate capacity to discriminate promising from unpromising properties (and magnitudes) to which to apply the principle of induction. Indeed we might generalize Quine's claim and argue that any principle of empirical reasoning, any form of ampliative inference, can only operate in the background of some favourable innate capacities or other. If so, we should *expect* that the Principle of Indifference will presuppose some innate capacity to select the 'natural' magnitudes to which to apply the principle. If all the indefinitely many magnitudes appeared equally natural, then the application of the Principle of Indifference would be as hopeless as the use of the Principle of Induction to generate functional generalizations.

Recently George Schlesinger has argued for a stronger conclusion, namely, that the Principle of Indifference is indispensable to all forms of inductive reasoning.[8] Schlesinger's argument does, I think raise an important question about the place of the Principle of Indifference in empirical reasoning, which can be argued to be far more extensive than suggested in the introduction to this essay. On the other hand, it is not clear that this argument, in part because of its brevity, establishes his major contention. Basically, until we have an account of just how the Principle of Indifference is integrated into empirical reasoning, it cannot be shown that it delivers the results Schlesinger claims for it.

Schlesinger's main example of the kind of problem in empirical reasoning that the Principle of Indifference solves is as follows. A soldier is required to select a gun for use in action from the pair (a) or (b). He is told that in tests (a) had misfired 2% of the time and (b) 5% of the time. It is plain that the rational choice is (a). But, Schlesinger says, how can this be for:

> The rifleman readily concedes that the commander is right that all the instances of the sample class have features in common which the instances yet to be encountered fail to exemplify. Furthermore, he acknowledges the total absence of positive evidence that these differences are of no significance, and may be ignored. But if we are not to ignore them, how should they feature in our deliberations? [9]

In this example, Schlesinger's question is what we do with the evidence that a sample differs in some way, (d), from the population but have no information of the significance, if any, for the frequency of some property (f) in the population. To hold that (d) would raise or lower (f) (i.e., the reliability of the gun) would be equally unfounded and hence arbitrary. This seems to mean that we rank equally that (f) increases reliability and that it decreases reliability. This, in turn, Schlesinger says, leaves the soldier with the reasonable belief that (a) will display 2% failure rate in the action and (b) a 5% failure rate in action.

Schlesinger holds that similar problems arise with other forms of empirical reasoning and a similar solution is appropriate:

> The Principle of Indifference plays a similar role in all cases of inductive reasoning, regardless of whether they lead to a singular or a general statement, to a probable conclusion or the confident enunciation of a law of nature. [10]

So even when, for example, the jury evaluates the evidence in a murder trial the same problem arises. Is it not possible, for all we know that some new evidence has just been found that totally overturns the seemingly watertight case of the prosecution? But, Schlesinger presumably insists, it is also possible that new evidence has just been found that really clinches the prosecution's case. Again while we cannot *ignore* these possibilities, we must treat them even-handedly, to avoid unfounded bias. Again this leaves the strength of the argument against the accused as it was. Now, however, we understand why the recognition of these possibilities does not upset our normal assessment.

Whether or not, Schlesinger's argument is correct, his examples bring out an important, if familiar, point. In adopting an epistemic stance, we must, in effect, evaluate both our evidence and our ignorance. Schlesinger's soldier has both his evidence of the sample tests and his ignorance of the relevance, if any, of certain conditions to reliability. We reason, as it were, from an island of knowledge in a sea of ignorance. Schlesinger is arguing that we are not entitled to reason from the island of knowledge (evidence) alone and ignore the ignorance. His aim is to provide a rational way to deal with the ignorance. If we accept his account of our epistemic situation, then we should draw the conclusion that in empirical reasoning we operate in part in conditions of evidence and in part in conditions of ignorance. If we accept the generality of Schlesinger's problem, then the conditions of ignorance that the Principle of Indifference operates in are not rare but ubiquitous. In any empirical reasoning, where there is evidence E for hypothesis H, there will also be a component of ignorance, of possibilities with respect to which we have no evidence, which must also be taken into consideration when determining our epistemic attitude to H. While this essay is not the place to pursue the connection, we might relate the foregoing discussion to the distinction between the support evidence gives to a hypothesis (how much the balance of evidence favours it over its negation) and the weight (or quantity) of evidence available. An increase in the weight of evidence is like an increase of the size of the 'island of evidence' and consequent diminution of the 'sea of ignorance'.

But while there is this important implication to Schlesinger's discussion should we accept the argument at the centre of his discussion? Has he shown how matters appropriately subject to ignorance "should feature in our deliberations"? In particular has he shown that the Principle of Indifference yields the kind of result embodied in our everyday thought — as illustrated by Schlesinger's case of the soldier selecting his gun? Should we claim that we should regard as equally probable that untested factors will increase as decrease reliability? If so, should we not recognize the third possibility that the factor will not affect reliability and regard this possibility as equiprobable with the other two? But since these three exhaust the possibilities, the Principle of Indifference yields the result that while the untested factor is no more likely to raise the reliability of the gun than to lower it, it is far more likely to change it (by lowering or raising it) than to leave it unaffected. But this, of course, is not the result Schlesinger sought. Hence while Schlesinger's line of argument properly brings out the importance of recognizing that we need rational responses to matters on which we are ignorant, and the Principle of Indifference is one such response, it is unclear how this principle, at least as Schlesinger deploys it, validates our natural responses to ignorance as a component of empirical reasoning.

University of Melbourne, Parkville, Victoria

NOTES

[1] Schlesinger (1991), Chapter 9.
[2] Salmon (1966), p.66.
[3] Resnik (1987), p.37.
[4] Rawls (1972), pp.152ff.
[5] Cf. Mellor (1971), Chapter 2.
[6] Salmon (1966), p.66.
[7] Wittgenstein (1964), section 149.
[8] Schlesinger (1991), pp.195-202.
[9] *ibid.*, p.200.
[10] *ibid.*

REFERENCES

Mellor, D. H. (1971) *The Matter of Chance*, Cambridge, Cambridge University Press.
Rawls, J. (1972) *A Theory of Justice*, Oxford, Oxford University Press.
Resnik, M. D. (1987) *Choices*, Minneapolis, University of Minnesota Press.
Salmon, W. (1966) *Foundations of Scientific Inference*, Pittsburgh, University of Pittsburgh Press.
Schlesinger, G. (1991) *The Sweep of Probability*, Indiana, University of Notre Dame Press.
Wittgenstein, L. (1964) *Remarks on the Foundations of Mathematics*, Oxford, Blackwell.

JOHN F. FOX

TOWARDS METAMETHODOLOGY: FOR THE HISTORY AND PHILOSOPHY OF SCIENCE

1. SCIENTISM

Much philosophy of science is methodology of science. How should one go about doing and evaluating it? The question is one of the methodology of methodology, i.e. of *metamethodology*. There is a vague thesis common to Descartes and more recent philosophers such as Quine and Lakatos: that what is good methodology, good evidence, good reason for accepting, rejecting or revising beliefs in mathematics and in the sciences properly so called, does not differ in significant kind from what is good methodology, evidence or reason elsewhere, even in epistemology and ethics; and further, that mathematics and the natural sciences provide good paradigms for methodology generally. Following von Hayek,[1] I call this form of epistemological and methodological monism: *Scientism*.

Scientism has been the inspiration for much of the philosophy of science; for studying how those who have conspicuously advanced knowledge have proceeded, hoping to distil some elixir of method or rationality to serve both as a physic for all ills of the mind, and as a tonic for its healthy use, in discovering and understanding truth. This can seem a foolhardy or fanatical hope. However, much more of the philosophy of science has been inspired by a *restricted* scientism: the idea that even if in some realms the methods by which the natural sciences achieved their spectacular successes prove applicable either to a limited extent or not at all, they have not yet been adequately exploited in some realms; in particular, in the philosophy of science itself.

Lakatos hinted at a point that deserves a label, which I shall call the *indifference thesis*: that what matters is to converge towards the whole truth about a domain, not whether one does this by ever less rash overstatement or by ever less timid understatement.[2] But there must be means for correcting rashness in the one case, timidity in the other. Testing for falsehood is a classical means for the former. For the latter, one can test for *inadequacy*[3] and push timid generalizations out towards their full domains.

From the indifference thesis we learn that to reap fully its heuristic benefit, one need neither affirm (unrestricted) scientism nor place *a priori* restrictions on it, but merely push it as far as one can. This illustrates another of Lakatos'

P.J. Riggs (ed.), Natural Kinds, Laws of Nature and Scientific Methodology, 103–121.

maxims, that one need not affirm or believe the hard core of one's programmes. So to our metamethodological question we have a preliminary answer: use the same methods used in good science.

Restricted or unrestricted, scientism by itself is relatively vacuous: it only generates a definite methodology when conjoined with a detailed account of just what, in the complex and evolving practices that have been called science, is deemed exemplary. After all, scientism does not hold up everything in the often greedy, conformist, authoritarian, status-crazy, cut-throat behaviour of scientists as exemplary. It is only their best scientific behaviour that it calls on us to emulate. Versions of scientism vary, as this best behaviour is variously identified.

2. WHEWELL'S METAMETHODOLOGY

William Whewell considered that it was a good thing to understand things and to make discoveries. He thought that natural scientists had shown themselves over a couple of centuries very good at this. He thought that if we wanted to learn how to understand things and to make discoveries, we should study how it had been successfully done, that is, study the history of the natural sciences. This would provide a basis on which a philosophy, a theory of method, could be built. So he produced the encyclopaedic *The History of the Inductive Sciences*, providing data for a comparably huge sequel, *The Philosophy of the Inductive Sciences*. Whewell is the very father of the history and philosophy of science.

It could seem that a key idea of Whewell's project was scientism. In fact it seems not to have been. His assumption was less "bold", more commonsensical and open-ended. He thought that there were some realms where the methods by which the natural sciences had achieved their spectacular successes would not be applicable.[4] But he also thought that many realms of learning in which such methods had not yet been exploited, including in particular the philosophy of the sciences, could and should learn from them quite straightforwardly. In short, he espoused restricted scientism. The indifference thesis reminds us that his lack of boldness was no vice.

Plato raised the problem of the seeker after truth. Only one who knew the truth would recognize it when s/he saw it. But to seek after truth is precisely to seek to know it. So to succeed in one's quest one must have already succeeded in it. His solution lay in his theory of recollection: one indeed knows the truth already, but it needs to be called to mind again. There is an analogous problem for Whewell. To discover how discoveries are made only makes sense if one does not yet know how to discover; but if one does not, how could one go about discovering how?

Another of Whewell's key ideas provides a partial answer to this platonic problem.[5] I call it, Whewellian bootstrapping, to distinguish it from Glymour's notion. (For brevity I shall take "Whewellian" as read, saying simply "bootstrapping".) We start out by using what, according to our best understanding, is good method. It need be no disaster if we start out with a crude, rudimentary and in some ways erroneous understanding of method. For we use this understanding in a way likely (by its own lights) to lead to its own improvement, and its replacement with a better.

Firstly, we continue to revise and improve our theory of method by continuing to test it against historical facts and improve its fit there. Suppose our initial method is crude Baconian induction: collect examples, form generalizations, test them against other examples, revise them if they are found wanting. We use this method to study the history of science, trying to find inductively the method that contributes to discovery and to increase in understanding. When we have generalizations about this, we test them against further facts from the history of science. If some generalization turns out not to fit the historical practice of great discoverers, we revise it.

Where the subject of investigation is method itself, there are further means of bootstrapping available. For secondly, we *use* our improved notions of method in our very enquiry. If, for instance, crude inductivism has been replaced as our best current theory of method, our subsequent study of the history of science should no longer be done in a crude inductivist way but in the best way we so far know how. If our revisions have been improvements, we are more likely to make the further discoveries about method that we desire. Thirdly, we can learn how to fine-tune methods not only from the history of science, but also from our own experience of using them.

3. WHY DID WHEWELL'S PROJECT LANGUISH?

Why, over the century after Whewell, when so many intelligent philosophers were impressed with science and concerned with its analysis, was Whewell's approach followed by so few? Christianity, according to Chesterton, was not tried and found wanting; it was found difficult and not tried. Whether that is so or not, Christianity has at least had nominal adherents in billions; Whewell's project, for over a century, did not attract them even in scores. Its difficulty is surely a partial, but just as surely an insufficient explanation.

Let's try a Sellarsian strategy and beat about some neighbouring bushes. Why did Huyghen's wave theory of light languish? Largely because there was a powerful rival that could explain a great deal that his theory could not, and which had inspired remarkable discoveries: Newton's corpuscular theory. Now an alternative approach managed to capture the allegiance of those who were attracted both to scientism and to empiricism. I suggest that the power and attractiveness of this approach largely explains the eclipse of Whewell's project.

Against Kant, it held that all knowledge was either analytic or strictly empirical. Analytic truths were necessary; their truth depended solely on the meanings of words, not at all on ways things happened to be, and our knowledge of them depended solely on our knowledge of such meanings, not at all on experience or observations. Our knowledge of empirical truths rested on experience or observations, and their truth was not necessary but depended on the way things happened to be. Methodology and epistemology, as traditionally understood, contained a host of confused questions which begged answers that were neither analytic nor empirical. Suitably reformed, they would contain only analytic or empirical matters. The analytic parts would typically comprise analyses of such concepts as "explanation", "law" or "reduction", and the principles of an analytic inductive logic; and the empirical would comprise a "naturalized" epistemology that was a "branch of empirical science", and which like such science would be "value-free". In a phrase, the rival philosophy that blotted out Whewell's project was (Logical) Positivism.

This picture in some realms did, and in many more was taken to, provide newly explanatory eludications and even anticipate startling discoveries in mathematics and the sciences; a list of headings will serve as reminders. After non-Euclidean geometries, geometry came to be seen as not *a priori* after all; after Dedekind, Frege, Peano, Russell and Whitehead arithmetic came to be seen as not being synthetic after all; the Special Theory of Relativity indicated that insisting on the empirical meaningfulness of notions led to spectacular progress; quantum mechanics suggested that the most advanced and basic physical theory required an instrumentalist rather than a realistic reading; subjectivist Bayesianism seemed to many at last to provide what earlier attempts (Carnap's inductive logic, Popper's sceptical falsificationism) had not, an account of scientific rationality in terms of the categories of empirical evidence and mathematically rigorous reasoning.

4. POSITIVISM VERSUS SCIENTISM

The commonest term for identifying the peculiar character of the method of the sciences to which they are taken above all to owe their success is "empirical". So if its philosophy wishes to mimic science, it should, it seems, strive to be empirical. But precisely the most strongly empiricist philosophies jib at this: for they typically hold that philosophy is in principle contrasted with the sciences precisely by lacking this peculiar character. This is largely because of their Kantian legacy. Though they reject synthetic *a priori* knowledge, they follow Kant in emphasizing the analytic/synthetic and *a priori*/empirical distinctions, and in treating the problem of understanding the success of science as a problem of *demarcation*: of explaining what is distinctive about science that makes such success possible there but not elsewhere.

Consistent scientism follows an older tradition, deriving from Descartes, who treated the problem by contrast as one of *method*. The problem was that of finding out by what means science has achieved what it has, in order to apply them in other realms to comparable effect. Thus, too, Whewell. These two approaches are incompatible. One seeks to explain why philosophy cannot, the other to learn how it can, emulate the merits of science.

Even demarcationist parts of the empiricist tradition show the pull of scientism in their accounts of philosophy. Among serious philosophers of science, Popper is a conspicuous example. Among others, many "linguistic philosophers" insisted in the first instance that philosophy, the analysis of meanings, was worlds apart from science, the study of facts, but tended in the second instance to characterize philosophy as a kind of empirical lexicography, and practise it by appeal to supposed facts about usage.

Like ethics, methodology and epistemology involve irreducibly evaluative predicates (e.g. "justified,", "unreasonable"), deem some things better than others. If such normative judgments are not adequately grounded simply by reports of sense-experience and logic, if science is empiricist, and if empiricism implies that only sense-experience and logical reasoning provide good reason or evidence, even restricted scientism is doomed.

Since the history of science shows compellingly that science is not empiricist in this sense, this point is academic. However, it is widely held that one aspect of the empirical character of the natural sciences that crucially helps explain their objectivity is their *value-freedom*; and even this spells trouble for restricted scientism. For those who espouse both scientism and some such strong form of empiricism must in consistency abandon epistemology and methodology. Many of them have recognized this, but fewer have been prepared to do so in so many words; some found alternative words. For one could of course still study in empirical, value-free fashion what methods scientists in fact used, what norms were in fact operative, what procedures were in fact generally considered sound or rational. Such study could be more a matter of sociology of science,[6] or of lexicography,[7] or even of behaviourist psychology.[8] It was suggested that those who had been addressing the bad old questions should eschew them, and devote themselves instead to the new; and it was suggested that these constituted a "naturalized epistemology".

Why this extraordinary co-option of the name of a supposedly discredited discipline for something completely different? Was it to disguise the fact that those who took the new path were no longer doing the jobs they had been appointed to do? But this is too simple. It was thought that such factual study was a traditional *part* of what had been done in epistemology and that it constituted the respectable and salvageable part. To some extent this was part of what had been (though seldom systematically) done. But *why* was it such a part? This was problematic. For how could *facts* about the history of science count as evidence for *value-judgments* of rationality, help settle questions of normative epistemology? In his review of Lakatos and Musgrave's *Criticism and the Growth of*

Knowledge, Jack Smart[9] expressed the objection of any consistent orthodox epistemologist. In effect it was this: what are these people doing citing facts from the history of science when they are arguing to methodological conclusions? Haven't they read Hume? Don't they realize that you cannot logically argue from an 'is' to an 'ought' ?

5. EMPIRICISM VERSUS POSITIVISM

In its more strictly empiricist versions positivism was riddled with problems of reflexive incoherence. Its own standards of warrant were, by those standards themselves, not warranted; its own standards of meaningfulness were, by those standards themselves, not meaningful. When scientism and empirical facts about science were taken into account, its problems multiplied.

Despite the orthodoxy that judgments of fact have no logical bearing on those of value, that philosophies of science fitted the practice of what was recognized as great science was almost invariably taken as a strong count in their favour. Einstein was widely thought to have discovered his theory of relativity by rigorously attending to the conditions under which such assertions as "A occurred at the same time as B" and "System K is in a state of absolute rest" could be empirically decided, and by either dismissing as meaningless or stipulating by convention what could not be so decided, e.g. the one-way speed of light. This was counted in favour of verificationism even by those who considered other, *a priori*, arguments to be decisive in its favour; even by those who considered other arguments to be decisive against it.

Still more, that philosophies of science *clashed* with what was recognized as great science was taken as a strong count *against* them: especially when other people's philosophies so clashed. If some account of rationality turned out to deem that the procedures by which Galileo, Newton, Lavoisier, Darwin, Maxwell, Rutherford, Curie, Einstein, Bohr and Heisenberg made their scientific breakthroughs were all irrational; then even the staunchest upholder of the logical chasm between facts and values was likely to doubt the adequacy of the account. Feyerabend used such clashes against all methodologies,[10] Lakatos against all others than his own.[11] Even Popper was prepared to use it. One of his central arguments against inductivists was that if they were right about rationality, scientists should prefer the more probable to the less probable, and so should extrapolate as little as possible beyond the observed facts rather than accept theories of wide explanatory scope, and should prefer, when confronted with unexpected difficulties, timid and *ad hoc* modifications of their theories, rather than bold new explanatory conjectures — contrary to the practice of the best scientists.

Could these philosophies coherently recognize facts from the history of science as evidence for and against philosophies? It was not clear that they could.

Positivistic philosophies of science tended to internal breakdown, not least because their scientism could not be squared with their extreme empiricism; as indeed had been suggested by Hume's problem of induction. This internal breakdown, and the influence of writers like Kuhn who criticized such philosophies on the basis of the history of science, made the relevance of history increasingly obvious. Joseph Agassi,[12] Imre Lakatos[13] and Paul Feyerabend[14] contributed partial accounts of just how it was relevant.

A central theme of Agassi (1963) was that methodologies could and should be evaluated with reference to the history of science. Each philosophy of science, he thought, generated its own style of historiography: inductivists wrote the history of science as a history of observations and inductions, singled out as problematic episodes that lent themselves with least facility to such reconstruction, and attributed error to careless observation or faulty reasoning. Analogously for other philosophies. The adequacy of the philosophy could be tested by the adequacy of the corresponding history. The logic of this is simple but deserves to be brought to light, if only because "the virtue of a logical proof is ... that it suggests doubts".[15] The philosophies in question are theories of rational method. What they directly say is "To be rational is to proceed thus." Agassi assumes that they lead naturally to writing histories of science which say "Science has overall proceeded thus." Clearly the tacit link is this assumption: "Science has overall proceeded rationally".

6. LAKATOS' METAMETHODOLOGY

Lakatos (1971) was engaged in a complex dialectic. One thing he was trying to do was to generalize a philosophy of science into a general theory of rationality, taking as a starting point Popper's account of science. This took "knowledge" to consist of conjectural theories, tested against basic statements (roughly, singular observation-reports), but as yet unrefuted. In the case of a clash, theories were to be revised so as to maximize growth of such "knowledge". To apply this philosophy to mathematics in his (1963-4) Lakatos needed an analogue there of Popper's "basic statements", and remarkably and fruitfully, chose *theorems*. Likewise, to apply the philosophy to methodology itself he required such an analogue, and chose what he called "normative basic statements", "evaluated particular historical claims about science" — e.g., "Planck's radiation constant was arbitrary".[16]

I have suggested elsewhere[17] that sense is best made of his arguments if we attribute to him what I call the Maxim of Strong Coherence, that whatever an epistemology says about the relation of scientific theory to basic statements, it should say about the relation of epistemology to its normative basic statements. This seems the basis of his argument against Popperianism[18] and other rivals to his own "methodology of scientific research programmes".

How, according to Lakatos, do facts from the history of science count as evidence for or against theories of rationality? Though it was evaluated facts he claimed were the basic statements, the facts that he in fact presented as crucial evidence were brute relative to such evaluation. They were either sociological facts about what theories or research programmes were accepted or preferred at a given time, or facts about what scientific facts were known at a given time. This is puzzling; for what related in an obvious logical way to his theory were highly evaluated facts, such as that one programme was progressive or degenerating relative to another.

However, there is logic in this. The thrust of Lakatosian case histories is always to display science as rational by Lakatosian standards. Some try to explain Copernicus' programme superseding Ptolemy's,[19] or Einstein's superseding Lorentz's.[20] The pattern of explanation is that the new programme was more progressive, because it had generated novel facts. Other case histories argue that rival methodologies have to deem scientists irrational, in holding out against e.g. Young's wave theory[21] or statistical thermo-dynamics.[22] The pattern of argument is that they were not irrational: the theories were not yet progressive relative to their older rivals, because such novel facts were not yet known. This is how facts about what facts are known at a time and about what programmes are preferred at a time are relevant to Lakatos' story.

Notice, however, that they are *only* relevant on the assumption that *scientists are generally rational* in their preferences among research programmes. Lakatos owes this idea and the idea that the associated historiography provides the crucial test for a methodology to Agassi (1963). It is an extraordinarily strong assumption, not to be taken as a basic principle of epistemology. Offered the alternatives of denying the relevance of the history and making it an absolute dictator, we should choose a third.

7. METHOD IN METHODOLOGY

Methodology is done, even methodically, but usually not with explicitly articulated or self-conscious method. There is no orthodox metamethodology; the question is simply rarely considered. The dominant methodology of science nowadays, i.e. theory of method, is the Bayesian.[23] Its method is usually *a priorist*. It is defended by appeals to the intuitive rationality of probability theory as a calculus of degrees of belief (or betting quotients) and specifically by appeals to theorems about its necessity for avoiding various 'dutch books'.

However Grunbaum,[24] Jeffrey[25] and Rosencrantz,[26] in their 1970s Bayesian counter-attacks on Popper, moved into the realm of appeal to the history of science. They did this, in the first instance, in the course of arguing that Bayesians could account for the rationality of actual scientific preference as well as and indeed better than Popper. Since then there have been several Bayesian reconstructions of episodes from the history of science, by e.g. Dorling,[27] Franklin and Howson,[28] Thomason.[29] These were intended to display both the rationality of those episodes and that the Bayesian apparatus adequately accounted for such rationality.

This is a small concession to fact. It is consistent to argue that the evidence for Bayesian principles is *a priori*, that they merely happen also to show their fruitfulness by illuminating those episodes in the history of science that happen to be rational; as it can been argued that the evidence for arithmetic is *a priori*, but that it happens also to show its fruitfulness by illuminating such empirical situations as arrays of pebbles or cuisenaire rods. The appeal to fact can consistently be limited to rebutting Popperian (or Feyerabendian) claims that no inductive methodology (or no methodology at all) can display certain paradigmatic scientific achievements as rational, and not counted as positive evidence for the methodology. But there can be quasi-Lakatosian defences of Bayesianism, arguing that it makes the best sense of historical science.

8. THE PRIMACY OF PRACTICE

I suggested that if some account of rationality turned out to deem irrational all the procedures or beliefs of the great scientific pioneers, the staunchest upholders of a chasm between facts and values would doubt the adequacy of the account. It is surely simply rational to take facts about the history of science as significant evidence for evaluating accounts of rationality and of methodology. With Agassi and Lakatos, I take some claims expressing evaluations of historical episodes as relatively hard facts, discrediting general evaluative theories that contradict them.

Historically, reflection on such historical facts has provided a crucial source of much impressive methodology. Aristotle's logic arose from reflection on the principles implicit in the practice of geometrical proof, Pappus' heuristic from reflection on the practice of devising proofs and solving problems, Descartes' general rules on method from reflection on his solution of crucial problems in mathematics and philosophy. Whewell's methodology, notably his pioneering discussions of colligation and consilience,[30] arose even more from reflection on the development of many disciplines, from astronomy and mechanics to geology and the theory of heat, than from his reflection on earlier writers on methodology, from Bacon to Kant.

All this suggests the *primacy of practice*: that it is more importantly practised method that gives rise to methodology than methodology that teaches method. Tempted by the stark black on white of aphorism, Hegel merely overstated the case. Forms of life need not have come to an end when philosophy paints its grey on grey; but they must have at least evolved a fair way. The owl of Minerva can spread her wings before dusk, but not before late morning.

Agassi's and then Lakatos' general accounts of how to compare methodologies deemed those accounts better which reconstructed more of what is generally called "science" as rational. This I reject. It does not take much study of the history and sociology of this to recognize that to deem almost all of it rational is a vast exaggeration, and that to *assume* this is sheer abdication of a much needed critical stance, a critical stance typical of many of the best scientists.

So I agree with one of Lakatos' suggestions and disagree with another. But can we do better than thus announce intuitions piecemeal? Is there some rational and systematic way of doing metamethodology? I suggest that there is. Lakatos took Popper's methodology, plus restricted scientism in the form of the maxim of strong coherence, to generate an initial metamethodology with an inbuilt strategy for self-improvement. This in turn, he argued, led dialectically to his methodology of scientific research programmes and the corresponding metamethodology (of "historiographic research programmes"). I suggest that we start rather with restricted scientism plus Whewell's metamethodology. It too has an inbuilt strategy for self-improvement, but has advantages over Lakatos' approach: it involves heuristic, not merely verdicts on programmes already developed and tried; it involves no absurdly strong science-worship.

9. SUGGESTED INITIAL REVISIONS OF WHEWELL'S PROJECT

If several ways in which a position needs revision are already clear, we do not need, in order to revise it, to await later historical research. There are serious problems with the project as I have formulated it so far. Here are a few.

Trivially: only what has not been produced before counts as a discovery. An adequate literature search commonly shows that what one thinks one has discovered one has at best rediscovered. Most putative discoveries are merely tentatively hoped to be discoveries, pending such search. What we can hope to develop skills for is at most the kind of intellectual work which, if its results turn out to be original, would count as discovery.

A candidate for a better phrase is "problem-solving". This is useful in the first instance for the number of new questions it readily brings to mind. May not problem-solvings fall into quite different classes which have to be tackled quite differently? May not the skills of solving problems in mathematics conspicuously tend not to carry over into those of solving problems in everyday life? May not the skills required to solve the kinds of "puzzles" that characterize Kuhnian "normal science", those of competent "hacks", be very different from the skills required to solve the problems that arise when a paradigm is in crisis, those of inspired "cranks" who refuse to be bound by the "normally" unquestioned framework assumptions of a discipline? Are not the skills involved in solving a problem by following an algorithm often different from those involved in solving a problem by devising a novel kind of algorithm? Taking "problem-solving" as the desideratum also suggests that the history of the traditional natural sciences is too narrow a field of paradigms. Even the most scientistic should recognize that to develop an adequate general methodology, the art and history of computer programming, for instance, might be comparably useful.

To embrace even a restricted scientism is to believe that an appropriately sophisticated and rational method would be likely to produce impressive results in all manner of fields other than the physical sciences. If at a time it is principally the latter that provide unquestioned paradigms of such method, this is mere historical contingency. Even in the time of Descartes and Spinoza, it was rather geometry that they took to provide such paradigms. We must be open to new kinds of paradigms.

Even taking the metaphor of "discovery" seriously, there is reason to doubt that a straightforwardly inductive study of the history of science is good method. What kinds of *objects* have been called "discovered"? People are said to have discovered gold and continents. If we are searching for gold, or vast new areas to colonize, studying how gold was found in Victoria and California, or how Newfoundland and Australia were reached by Cabot and Cook, may be of interest and even of use; but copying them is likely to be as ineffectual as Don Quixote copying the heroes of the literature of chivalry. For there is reason to believe that the world has changed: that the planet contains no more continents still undiscovered by speakers of European languages, and vastly fewer large nuggets of gold visible on the surface of the earth. Perhaps there is much gold still to be discovered, but if so, more sophisticated methods are required, and indeed available, than first did the trick in Victoria and California. Perhaps there are vast habitable continents still to be discovered, but space travel and all its intricate heuristic may be needed for that. This moral was again available from Lakatos (1963-4): methods that were historically crucial to particular bits of progress may be already clearly superseded.

This suggests another obvious perspective on the platonic problem. After all, there is an analogous problem in science. We are trying to acquire knowledge about the world; to do this we must use knowledge about the world, and so already to have it. This sounds more like a sophism than a paradox, for obviously the knowledge we use is different from the knowledge we are trying to acquire — it is knowledge of different things. Describing them both as "knowledge of the world" merely obscures their differences by finding a vague and general common description. Analogously, to learn how to discover, one must already have discovered: but different things. To know how to make [some] discoveries is not the same as to know how to make [all] discoveries; just as to know some things is not to know all, and to know how to make some things is not to know how to make all.

Paradox lost becomes paradox regained only if there emerges some good reason to think that "knowing how to make discoveries" names only one skill or item of knowledge. Once articulated, this is obviously wrong. A main virtue not merely of logical proof, but of explicit articulation, is that it suggests doubts; this was a refrain of Popper's. But this seems to have been a constitutive assumption of Whewell's project.

Still, like scientism itself, or theories like Newton's of gravity or Maxwell's of the ether, a project may be very fruitful despite resting on erroneous assumptions, and the search for general heuristic is I think in this category. Descartes created analytic geometry only by rejecting the dogma that different subject-matters (number and space) required different methods of study. Lakatos (1963-4) was also inspired by belief in a thoroughly general heuristic. The belief may have been false. Later he thought it was, and followed Kuhn to the opposite extreme, deeming heuristic, positive or negative, to be entirely internal to each programme. This may also be false; it is surely less fruitful. As with scientism, as with Newtonian theory, the route to take is to push the quest as far as we can, to develop as rich a general account of how to discover as we can: we best learn the limits of the project by trying to surmount them. This is a better moral to be drawn from Kuhn.

10. THE JESUIT RULE

A rule of the Jesuits reads: "If there is some difference of opinion among us, and it seems that it should be aired, reasons should be adduced modestly and charitably, in the spirit that truth should have its rightful place, not that they should appear superior in the matter."[31] Arguably, Ignatius did not mean this to be of universal application. Two restrictions are implicit. First, "we" were the Jesuits; unity and charity among whom was of importance, for the effectiveness of their work, assumed to be Christ's. But that work could involve being very polemical against those who were (knowingly or not; for those who were not with Him were against Him) on the other side. Second, the difference mentioned is one of opinion: where the truth was known, or a matter of faith, differences were not considered those of *opinion*. But perhaps the rule should be extended beyond these restrictions; and Loyola was not saying that it is all right to be concerned with seeming superior rather than for the truth, so long as the opponents were heretics. How would it fare as a rule for all humanity?

Scientism assumes that the ways progress has been made in the sciences display the best ways of using our minds; should, indeed, be copied by those of us who wish comparably to advance mankind's understanding and mastery of the universe. As a matter of historical record, many who contributed spectacularly to the growth of science were no observers of the maxim, but obsessively concerned with their own status, extremely conceited, reluctant to take the most obvious points made by rivals and to acknowledge them if taken, happy to distort the history of research in a way that aggrandized themselves, their party or country, given to denigrating others unfairly. Many other genuine if less conspicuous contributors, as Kuhn (approvingly!) emphasized, refuse to entertain any ideas opposed to those they have been brought up on. So much scientism would hold the Jesuits' maxim in scorn. But so what? Like other mortals, competent scientists have virtues and vices; surely it is their virtues, not their vices, that we should be emulating.[32] This reply is obvious enough to be a truism, yet relevant enough to be devastating. The Jesuit rule can ground a strong argument against uncritical scientism.

Truisms are not by themselves loadbearing. To be content with such a reply is to be too complacent and unquestioning, to assume that we know all that needs to be known about intellectual ethics, and are virtuous compared with the scientific mob who (oddly enough) have been thinking to some impressive effect. Better a fruitful error than a barren truth, said Bacon, and in this bold aphorism was the seed of the whole new methodology of science. Better a fruitful vice than a barren virtue, implied the economic philosophers of the 18th century, and redefined morality so as to christen capitalism. Under the same banner, one could (it was Lakatos' strategy) reshape intellectual ethics in mimicry of successful scientists. Why not?

Can we do better in this conflict than merely take sides? In several ways; here is one. If we are to revise the standards of intellectual ethics on the basis of the success of scientists, we should investigate carefully and rigorously just what aspects of their practice help *explain* this success. This motivates taking the history and sociology of science very seriously — and for demanding there higher standards than are yet customary. Even where such explanatory relevance is evident, there may be adequate reason not to take the practice in question as normative: reason, e.g., to think that circumstances have changed in such a way that the practice is no longer likely to lead to success, or that a different practice is even more likely to lead to success.[33]

11. FALLIBILISM

With the positivists' extreme interpretation of "empirical", the thesis that science was empirical was not historically tenable. Yet science is surely a *broadly* empirical enterprise and very different from an a priori one. I take science to be generally *fallible* in ways that are relevant to scientism. Discovery of new facts, for example, can practically force us to revise general beliefs we have taken as knowledge — e.g. that all swans were white. But discovery of new theories can also lead us to revise factual beliefs we have taken as knowledge — e.g. that the sun sinks in the west. The fool on the hill sees the sun going down, but the eyes in his head see the world spinning round.[34] Given the traditional usage in which "knowledge" implies "truth", scientific "knowledge", whether factual or theoretical, may turn out not to have been knowledge after all.

Theoretical discoveries in science often require correcting or modifying not only what had previously been accepted, but the very realm of data they explain, explaining which indeed sometimes constitutes almost all of the evidence for them. A paradigm of this is the way Newton's theory of gravity required revising both Kepler's laws of planetary motion and Galileo's law of free fall.

Feyerabend puts the point generally. He uses the terms "reason" and "practice" as correlatives for e.g. chess theory and chess playing, logic and geometry, philosophy of science and chemistry, training long-distance runners and doing the running. He emphasizes that both sides of the contrasts are practices in the sense of things done, and that neither are irrational in the sense of void of reasons or rationale. The contrast is simply that the first of each pair involves explicit reflections on the rationale, and theorizing about the best way, of doing the second; involves *epistemic ascent.*

I suggested in §8 that by and large "practice" has primacy over "reason". But neither practice is infallible, or has invariable primacy. Sometimes chess theory shows what is wrong with certain chess practice, sometimes the other way round.[35] On this Feyerabend, as often, was sane and judicious under his jester's garb. On the same issue Lakatos, as often, flipped between extremes while parading in the robes of reason: in one breath announcing rules for the rejection of entire research programmes[36] under objectively defined conditions,[37] and suggesting a privileged place for philosophers of science in their policing;[38] in the next claiming it was intolerable hubris for a philosopher to give "advanced" sciences any advice at all,[39] and proclaiming this an "all-important" bit of progress in the theory of rationality.[40]

The point that both factual and theoretical judgments are liable to revision can tentatively be generalized. There seems, in science, to be no readily ascertainable category of statement or assumption that is immune from revision. That is, if any category is nominated ("established facts", "truths" or whatever) then there will be claims which would be judged at a time, according to ordinary or generally accepted criteria, to belong to that category, which at a later time would be generally agreed to have been revised rationally.

Despite each element being subject to possible query and modification, science at a time is cumulative, in that a new discovery rests on taking as given many old discoveries. As one might expect, this phenomenon becomes more marked with time — that is, discoveries tend to depend on more and more old ones. In science, important "discoveries" which have served as part of the basis for a host of further ones may have to be jettisoned themselves, yet leaving what has been built on them largely secure (though subject to some, perhaps quite drastic, theoretical reinterpretation). Vast amounts of fairly accurate astronomical information were built up on the basis of the assumption that the stars were fixed on a solid sphere that revolved daily around the earth. The theory that there were atoms (indivisible minima) of all the chemical elements was the basis for most of the chemical theory of the 19th century, and the assumption that the universe was pervaded by a substance of which light was a vibratory motion was the basis for the development of electromagnetic theory.

If nothing in science is unrevisable yet everything in science rests on other parts of science, one simply has to start where one is. The classical image for this is Neurath's: we are rebuilding our boat at sea, and while we do so, we have to rest on parts of it which may later need replacing in turn. This puts the "cumulativeness" of science in a less "foundational" perspective. But how can we tell that what we are using is an advance, or that our initial methods or factual assumptions are likely to lead to advances rather than regress? I think we can't tell for sure. We *hope* that the perceptual and methodological propensities biology and culture have provided us with are not too awry.

Some think that the survival of our species so far is proof that we have rationality mainly right. But many species that survived much longer than ours survive no more. Ours has already devised, extravagantly funded and based national policies on "theories of rationality" that dictate readiness to wage nuclear war. Both Popperianism and subjectivist Bayesianism imply that one simply has to start from where one is at first, and that one's attempts to improve matters are guesswork; on this point I concur with both. There is warrant for attributing to our fallible starting-points neither certainty nor high objective epistemic probability.

12. EXPLAINING PROGRESS AND CONSENSUS

Science has for centuries been a byword for progress and agreement. This has been as much a matter of rumour as of fact; historians from Duhem through Kuhn have emphasized the huge and enduring disagreements among scientists. The "internal" factors cited in time-honoured fashion to explain such progress and consensus are observation-reports certified by mere experience, and logic certified by mere understanding of meanings. Fallibilism rejects this explanation. Indeed, to the limited extent that such progress and consensus has been a matter of fact rather than rumour, much of the explanation lies in a host of "external" sociological factors.

On what central assumptions does the Whewellian project, of learning good method from a more or less scientific study of the sciences, rest? On at least two: that such study is likely to lead to a clear and detailed explanation of how the sciences have achieved their progress and consensus; and that this will be largely in terms of their method. Does not the "concession" that the explanation will be largely in terms of "external" factors undermine the entire project? If an adequate external account of some or most episodes in the sciences can be provided, this could discredit the idea that method has a significant role in their explanation; particular internal and external explanations can be rivals. But in general the internal and the external are complementary aspects of an overall explanation.[41]

Several paradigmatically "internal" factors do seem crucial to the explanation. One is a general policy of accepting, in the first instance, *new observations* (or very many such — the question of the conditions under which they are not even tentatively accepted is too complex to explore here), which can create tension among accepted beliefs. What has been discredited is not that observations are relevant, but that they are rendered *certain* by experience. Though the standards for acceptable observation reports change over time, that there generally are at a time such agreed standards (allowing for borderlines) helps explain such consensus as there is. This is, I think, the core of truth in an empiricist account of science; and why Popper and I can both think of ourselves as somehow empiricists.

Another is the use of logic to identify clashes and deem them unacceptable. I include the inferential "practice", not just the formalization that counts as the correlative "reason". Such "logic" includes much broad intuitive mathematics (e.g. calculus) and a theory of acceptable experimental error. All of this has changed over time; standards of acceptable approximation, in particular, are extremely relative to context and resources. The doctrine that has been shaken is that logic is rendered certain *a priori* by inspection of meanings. But agreement at a time about logic also helps explain such consensus as there is.

A third factor, not recognized in the heyday of the positivism that tried to reduce methodology to logic is widely agreed manners or methods for (e.g.) *resolving* clashes, evaluating and testing conjectures etc. Historically, despite valiant efforts like Whewell's and Popper's, this has typically not been developed into explicit methodology, but has been left at the level of Duhem's vague "educated good sense". So three "internal" categories, of observation, logic and method, are all crucial.

No matter how adequate the explanations that may emerge in terms of such internal factors, there will be external aspects to the explanations. Consensus on, e.g. language, procedures and the prestige of various authorities will be factors involved at every level and much of the explanation of all of this will be paradigmatically social. Also, even and especially the most internalist explanations are ideal-type explanations and so typically require supplementation with classically externalist explanations in order accurately to fit the facts. The need for external explanations precisely in such "Galilean" reconstructions makes dubious the opposition between the two proclaimed in the rhetoric of Lakatosians and supporters of the strong programme.[42]

I suggest that the quest for internal explanations, like that for Newtonian explanations, and scientism in general, is something to be pushed as far as possible, in order to find out how far it can go. If it can go a long way and shed a great deal of light, so much the better; if it quickly emerges how drastic are its limitations and where they lie, so much the better; only by the pushing are we likely to learn either.

13. APPLICATION TO METHODOLOGY

In methodology too discoveries typically require correcting or modifying what had been previously accepted, through Whewellian bootstrapping. But just we must start with the revisable statements we get by accepting observation reports laden with the theoretical and cultural assumptions of our language, time and communities, we must start with, and so take as good enough in the first instance, the methods we have, however crude.[43] Logic can be presumed common to science and methodology. So, *mutatis mutandis*, I suggest after Whewell and Lakatos, can the use of methods — e.g. of revising in the face of anomaly - that are intersubjectively recognized as rational.

More problematic is the analogue in methodology of the third internal element I mentioned, accepting new observations.[44] Space is scarce, so here I will say just this. We need not wait for an adequate treatment of the problem in order to proceed. Practice is primary. Just as in science we have been able to make great progress without having to wait on theoretical resolution of the problems raised by the theory-ladenness of observation statements, so we can in practice recognize with tentative confidence cases of scientific achievement, progress, discovery, problem-solution. That is enough to get the Whewellian project going.

La Trobe University, Bundoora, Victoria

NOTES

[1] von Hayek (1955). My characterization differs from his in three ways; he was primarily characterizing a practice rather than its rationale; he referred only to the natural sciences and not to mathematics; and building in his disapproval, he characterized the imitation of their methods as "slavish".

[2] Lakatos (1976). In fact "Beta" did not speak of "the whole truth", of rashness or timidity, but of ever false over-statements and ever true understatements, (p.58) and "Mu" pointed out the openness of most questions about such convergence (p.64). These are of course not the only two possible serious methodological strategies: Popper recommends ever *more* bold overstatements, so long as they survive rigorous testing.

[3] His "*heuristic* counterexamples" (p.83) function in such tests as "logical counter-examples" do in tests for falsehood. Cf. Fox (1981) p.93.

[4] Whewell (1847).

[5] For a further partial answer see §9 below.

[6] That epistemology is dead and that the empirical sociology of science is its sole legitimate successor is the thesis not of an individual but of a current subdiscipline, "social studies of science".

[7] Cf. Goodman's celebrated (1955) dissolution of the old and proclamation of the new "riddle of induction". See also the discussion by Clendinnen (this volume).

[8] Cf. Quine (1969).
[9] Smart (1972).
[10] Feyerabend (1974).
[11] Lakatos (1971).
[12] Agassi (1963).
[13] Lakatos (1971).
[14] Feyerabend (1978).
[15] Forder (1927) p.viii.
[16] Lakatos (1971) p.133 n.4 (see also p.115).
[17] Fox (1981).
[18] Cf. Lakatos (1971) p.124: "This meta-methodological rule [that "a rationality theory is to be rejected if it is inconsistent with an accepted 'basic value judgement' of the scientific élite"] would seem to correspond to Popper's methodological rule — that a scientific theory is to be rejected if it is inconsistent with an ('empirical') basic statement".
[19] Lakatos & Zahar (1976).
[20] Zahar (1973).
[21] Worrall (1976).
[22] Clark (1976).
[23] It is certainly not the dominant *method.*
[24] Grunbaum (1976a), (1976b), (1976c).
[25] Jeffrey (1975).
[26] Rosenkrantz (1977).
[27] Dorling (1979).
[28] Franklin & Howson (1984).
[29] Thomason (1994).
[30] For a good recent discussions see Forster (1988).
[31] Common Rule 26, *Rules of the Society of Jesus*, Rome 1935.
[32] Cf. §1 above.
[33] Cf. §9 above.
[34] Lennon & McCartney, "The Fool on the Hill".
[35] Feyerabend (1978).
[36] Lakatos (1978) Vol. 1 p.112.
[37] Lakatos (1978) Vol. 1 p.90.
[38] Lakatos (1978) Vol. 2 p.228.
[39] Lakatos (1978) Vol. 1 p.137.
[40] Lakatos (1978) Vol. 1 p.103n.
[41] Cf. below.
[42] This is well shown by Koertge (1976).
[43] See §2 above.
[44] See §6 above.

REFERENCES

Agassi, J. (1963) *Towards an Historiography of Science*, Beiheft 2 to *History and Theory*. Reprinted by Wesleyan University Press, Middletown, 1967.

Clark, P. (1976) 'Atomism versus Thermodynamics' in Howson: 41-105.

Dorling, J. (1979) 'Bayesian Personalism, the Methodology of Scientific Research Programmes, and Duhem's Problem', *Studies in the History and Philosophy of Science* **10**: 117-87.

Feyerabend, P.K. (1974) *Against Method.* New Left Books, London.

Feyerabend, P.K. (1978) *Science in a Free Society.* New Left Books, London; Part One, "Reason and Practice".

Forder, H.G. (1927) *The Foundations of Euclidean Geometry.* Dover, New York, 1958.

Forster, M. (1988) 'Unification, Explanation and the Composition of Causes in Newtonian Mechanics', *Studies in the History and Philosophy of Science* **19**: 55-101
Fox, J. (1981) 'Appraising Lakatos', Australasian Journal of Philosophy 51: 92-103.
Franklin, A. and Howson, C. (1984) 'Why do Scientists Prefer to Vary their Experiments?', *Studies in the History and Philosophy of Science* **15**: 51-62
Goodman, N. (1955) *Fact, Fiction and Forecast.* Harvard University Press, Cambridge Mass.
Grünbaum, A. (1976a) 'Is Falsifiability the Touchstone of Scientific Rationality?' in Cohen, Feyerabend & Wartofsky (eds.), *Essays in Memory of Imre Lakatos*: 213-52.
Grünbaum, A. (1976b) 'Is the Method of Bold Conjectures and Attempted Refutations Justifiably the Method of Science?', *British Journal for the Philosophy of Science* **27**: 105-136.
Grünbaum, A. (1976c) 'Ad Hoc Auxiliary Hypotheses and Falsificationism', *British Journal for the Philosophy of Science* **27**: 329-362.
Howson, C. (ed.) (1976) *Method and Appraisal in the Physical Sciences.* Cambridge University Press, Cambridge.
Jeffrey, R.C. (1975) 'Probability and Falsification: A Critique of the Popper Programme', *Synthese* **30**: 95-117; and 'Replies', *ibid.*: 149-57.
Koertge, N. (1976) 'Rational Reconstructions' in Cohen, R., Feyerabend, P.K. and Wartofsky, M. (eds.), *Essays in Memory of Imre Lakatos*: 359-369.
Lakatos, I. (1971) 'History of Science and its Rational Reconstructions' in Buck R.C. and Cohen, R.S. (eds.), Boston Studies in the Philosophy of Science 8, Reidel, Dordrecht: 91-135. Reprinted in Howson: 1-39 and in Lakatos (1978), Vol I: 102-38.
Lakatos, I. (1976) *Proof and Refutations*, Cambridge: Cambridge University Press 1976; Reprinted from (1963-4) 'Proofs and Refutations', *British Journal for the Philosophy of Science* 14-15.
Lakatos, I. (1978) *Philosophical Papers*, 2 Vols, Cambridge, Cambridge University Press.
Lakatos, I. and Zahar, E. (1976) 'Why Did Copernicus' Programme Supersede Ptolemy's?' in Westman, R. (ed.) *The Copernican Achievement*: 354-383. Reprinted as ch. 5 of Lakatos (1978) Vol. 1.
Quine, W. (1969) 'Epistemology Naturalized' in his *Ontological Relativity and Other Essays.* Columbia University Press, New York.
Rosencrantz, R. (1977) *Inference, Method and Decision.* Reidel, Dordrecht; ch. 6, 'Bayes and Popper'.
Smart, J.J.C. (1972) 'Science, History and Methodology', *British Journal for the Philosophy of Science* **23**: 266-74
Thomason, N. (1994) 'The Power of ARCHED Hypotheses: Feyerabend's Galileo as a Closet Rationalist', *British Journal for the Philosophy of Science* **35**: 255-264
Urbach, P. (1989) 'The Bayesian Alternative to the Methodology of Scientific Research Programmes' in Gavroglou, K., Goudaroulis, Y. and Nicolacopoulos, P. (eds.), *Imre Lakatos and Theories of Scientific Change. Boston Studies in the Philosophy of Science* 111. Kluwer, Dordrecht.
von Hayek, F. (1955) 'Scientism and the Study of Society' in *The Counter-Revolution of Science.* Free Press, Glencoe, Ill.
Whewell, W. (1847) *Philosophy of the Inductive Sciences*, 2 vols., 2nd edn; Reprinted by Johnson Reprint Corp, New York, 1967.
Worrall, J. (1976) 'Thomas Young and the "Refutation" of Newtonian Optics' in Howson: 102-179.
Zahar, E. (1973) 'Why Did Einstein's Research Programme Supersede Lorentz's?', *British Journal for the Philosophy of Science* **28**: 195-213. Reprinted in Howson: 211-75.

NEIL THOMASON

ELK THEORIES – A GALILEAN STRATEGY FOR VALIDATING A NEW SCIENTIFIC DISCOVERY*

In January 1610, Galileo turned his telescope toward Jupiter and soon afterwards announced that he had discovered four moons circling it. Although a considerable controversy ensued, by the summer of 1611 a consensus had emerged among the knowledgeable that the telescope was reliable (at least in this case) in the super-lunar realm and that the Jovian system was real. This is a remarkably fast closure of serious debate about a complex scientific issue, considering the slowness of early 17th century communication, the problematic nature of the new instrument and the importance of the astronomical discovery.

Was the acceptance of the reality of Jupiter's moons and the reliability of the telescope a rational one? Or, as Feyerabend proposed, was it reasonable in 1612 to believe that the telescopic reports were of artifacts and that Galileo prevailed "... because of his style and clever techniques of persuasion ... and because he appeals to people who are temperamentally opposed to the old ideas and the standards of learning connected with them"?[1]

A crucial requirement for answering such questions is an independently justified criterion for distinguishing experimental artifacts from real phenomena. Yet, as we shall see below, the ways of doing this so far proposed by philosophers of science leave it unclear whether the acceptance of the Jovian moons was rational. However, we shall also see that Galileo had an illuminating argument which gives an additional way to establish that an observation is valid.

Certainly, the more popular discussions about the reliability of the telescope lasted much longer than the debates among the experimentally inclined. Over half a century later, in 1670, Henry Stubbes, "Physician at Warwick," published his eloquent, spirited attack on Glanville and the Royal Society, "*The Plus Ultra reduced to a Non Plus: Or, a Specimen of some* Animadversions *upon the* Plus Ultra *of Mr. Glanvill, wherein sundry* Errors *of some* Virtuosi *are discovered...*". The book includes a section "The Deceitfulness of Telescopes".

Samuel Butler, whose *Hudibras* was the most popular poem of the era, commented on this debate in his satirical "The Elephant in the Moon," in which members of a learned society turn their telescope toward the moon and "see" a battle. To their surprise and joy,

... A stranger sight appears
Than e'er was seen in all the spheres!

P.J. Riggs (ed.), Natural Kinds, Laws of Nature and Scientific Methodology, 123–144.

A wonder more unparallel'd
Than ever mortal tube beheld;
An Elephant from one of those
Two mighty armies is broke loose,
And with the horror of the fight
Appears amazed, and in a fright
(Butler 1671 ?/1854, ll. 121-28).

Seeing the "elephant" move extremely rapidly across the moon, one savant, with a spectacular series of transparent sophisms, proclaimed that they could "make this great phenomenon / (Were there no other) serve alone, / To clear the grand hypothesis / Of th' motion of the Earth from this." (ll 311-14) The Learned proclaim that their great "discoveries" will immensely help their social position:

[We are] henceforth to be believed,
And have no more our best designs,
Because they're ours, believed ill signs ...
This one discovery's enough,
To take all former scandals off
[ll. 200-2 & 205-6].

(In fact, "The Elephant in the Moon" is one of the first social constructivist analyses of science.) While the Learned are in a tizzy over their discovery, the footboys find that the armies are merely "swarms of flies and gnats" and the elephant is a mouse in the telescope.[2] A decade later, "... the English astronomer John Flamsteed still deemed it necessary to show by an analysis of lens systems that lenses and combinations of lenses 'doe not impose upon our senses.' "[3]

Philosophers' proposed ways of distinguishing experimental artifacts from real phenomena are not generally applicable to Galileo's problem with Jupiter. In some situations, there is no difficulty in showing that a new scientific instrument is reliable or that a posited new entity is real. If one has a well-supported theory of how the new instrument is to work, or if one has an old, reliable type of instrument that detects the phenomenon, then the epistemological problem is easily resolved. The situation is more challenging, however, when one's theories cannot explain how an instrument works and one cannot independently test its reliability. Galileo was in this more challenging situation — there was no independent way of determining the reliability of telescopic observations of Jupiter's moons and, as we will see in some detail below, Galileo did not have a well-supported optical theory of the telescope.

Nor did the moons do well on other strategies philosophers have proposed. For example, the moons hopelessly failed the spray test that made Hacking an entity realist: "As far as I'm concerned, if you can spray them then they are real."[4] They do no better on his more general criterion that whatever you can manipulate is real. Clearly they did not pass his immensely plausible independent confirmation test — that one can accept a result or an entity as real if one has measured it by two or more independent experimental techniques. But, Galileo had only one sort of experimental technique — the optical telescope. More ominously, he was pressing it to its limits — Jupiter's moons could only be seen by the best telescopes in fine viewing conditions.

In *Experiment Right or Wrong,* Franklin builds on Hacking's suggestions with an admittedly tentative list of nine strategies for distinguishing experimental artifact from valid observation. Here is Franklin's list of strategies that provide a reasonable belief in the validity of an experimental result:[5]

1. Experimental checks and calibration, in which the apparatus reproduces known phenomena.
2. Reproducing artifacts that are known in advance to be present.
3. Intervention, in which the experimenter manipulates the object under observation.
4. Independent confirmation using different experiments.
5. Elimination of plausible sources of error and alternative explanations of the result.
6. Using the results themselves to argue for their validity.
7. Using an independently well-corroborated theory of the phenomena to explain the results.
8. Using an apparatus based on a well-corroborated theory.
9. Using statistical arguments.

Below, generalizing an argument of Galileo's, I will propose a tenth strategy to be added to Franklin's list. Seven of Franklin's strategies were not available to Galileo: his telescope was not reproducing known artifacts; he did not manipulate the moons; they had no independent experimental confirmation; there was no systematic elimination of alternative explanations; there was not an independently well-corroborated theory of Jovian moons; there was not a well-corroborated theory of the telescope; and statistics played no role. However, Galileo did have access to Franklin's two remaining strategies of establishing that the observations were not artifacts. Below, I will discuss how one of these, naked-eye observations, could check telescopic observations in some areas. But first, we need to turn to the other strategy.[6] Franklin describes it this way:

> ... sometimes the results themselves may argue for their own validity. Consider Millikan's observations that the charges on oil drops were always integral multiples of a certain unit of charge. Although we might very well believe that Millikan's apparatus might give incorrect values of the charge on the drops, there was no remotely plausible malfunction of the apparatus that would give rise only to integral multiples of a single unit. Similarly, although one might believe, as Cremonini and other seventeenth century Aristotelians are reported to have be-

> lieved, that Galileo's telescope could create specks of light, it is extremely unlikely that it could create specks of light that behave as a planetary system with consistent motions and eclipses. It would be even more astonishing if such artifacts obeyed Kepler's Third Law, R^3/T^2 = Constant. Thus, one supported the validity of the observations of the moons of Jupiter.

Franklin acknowledges in a footnote that "This is not a historical argument, but rather one that could have been given." As we will see below, his argument is anachronistic, but his logical point is an important one. In many situations, scientists advance two or more independent and otherwise *ill-supported* hypotheses to explain a surprising phenomenon. One of these ill-supported hypotheses can be that the instrument that reported the phenomenon is reliable. Franklin's account of Millikan's implicit argument for the reliability of his oil drop apparatus is an excellent example of someone arguing in this way. It is worth discussing briefly, so that we might more vividly see both how Galileo's argument for the super-lunar reliability of the telescope relied in part on such reasoning and how, at one key point, he went beyond it in a particularly interesting way.[7]

Millikan is using what I elsewhere called ARCHED (Advance Reinforcing Chancy Hypotheses Explaining Data) Hypotheses.[8] There is a surprising regularity in the data. On an ARCHED Hypotheses analysis, this data is best accounted for by two independent and otherwise ill-supported hypotheses: there is a basic unit of electrical charge *and* Millikan's oil drop apparatus works well (i.e. no convection currents, etc.). That is, Millikan's claim to have established the minimal unit of electrical charge with his problematic apparatus is justified because the best explanation of the observed changes in the droplets' positions is explained by positing *both* that the apparatus is reliable *and* that there is a minimum unit of electrical charge. Neither chancy hypothesis by itself can explain the regularities seen in the apparatus. If the apparatus is reliable, but there is not a minimum charge, then one would probably not have seen the regularities in the change in position. Conversely, if there is a minimum electrical charge but the apparatus was not reliable, one also would probably not observe these regularities. Both chancy hypotheses conjoined are required to explain the regularities. Like the two sides of an arch, neither hypothesis stands supported by itself — each requires the other.

But, surprisingly, in certain conditions, even the demand for a second, independent hypothesis to form ARCHED Hypotheses is too stringent. In such conditions, one can show that an instrument is reliable even if there is no adequate account of what is observed, no adequate theory of how the instrument works, and no independent way of determining that the instrument is reliable. Whereas ARCHED Hypotheses require two independent problematic hypotheses for them to mutually support each other, in certain conditions the claim of an instrument's validity can stand alone supported by the instrument's unpredicted and unexplained output itself.

Put as I just put it, the claim that one can show that an instrument is reliable with such minimal evidence is surprising — certainly philosophers have normally demanded higher standards for establishing reliability. Yet, as we shall now see, there is another way of viewing this matter, wherein establishing that an instrument is reliable is a hum-drum fact of common sense. One can establish that an instrument is reliable when our understanding of how the instrument works, while grossly inadequate as a theory, makes it very unlikely that the observed output is an artifact. Of course, this depends on our views of what sort of artifacts the instrument is likely to produce. How could such views of the instrument be justified, in the absence of anything approaching an adequate scientific theory of how it works?

The key is that even a new instrument is rarely a Black Box, the workings of which are *completely* unknown. Rather, even in the absence of an adequate theory of how it works, we can give a (very) rough account of how it works. We can say that optical telescopes deflect or reflect light; linear accelerators somehow accelerate tiny particles to tremendous velocities and then run them into 'targets'; free magnets usually point in the same direction because of the Earth's magnetic field; etc. While these accounts are not nothing, they cannot be called 'theories' on any but the most expansive use of the term.

I will call such minimally contentful accounts "Elk Theories" in honour of John Cleese's Monty Python character. To the increasing horror of the highbrow television presenter, Miss Anne Elk lovingly repeats her ("That is my theory, it is mine, and belongs to me and I own it, and what it is too") theory of the brontosaurus:

My Theory
By A. Elk (Miss)
All brontosauruses are thin at one end, much much thicker in the middle and then thin again at the far end.

As the presenter rightly comments: "Well, er, this theory of yours appears to have hit the nail on the head."[9]

Elk Theories are far from rare — they permeate our lives. More importantly, despite their minimal content, Elk Theories of instruments are much more philosophically and practically useful than one might expect. An Elk Theory can give considerable confidence that there is only a certain range of artifacts that the instrument itself can produce. Although the boundaries of this range are quite vague, there are many cases that clearly fall outside that range. If the instrument produces results that the Elk Theory of the instrument says probably are not artifacts, one can be pretty confident that the instrument is being reliable. Below, we will see that a key Galilean argument for the extra-terrestrial reliability of the telescope centrally relies on an Elk Theory of the telescope.

First, it is worth reflecting on how surprisingly minimal an Elk Theory need be for establishing an instrument's reliability. A simple case is litmus paper. Litmus paper has some chemical in it that makes it turn pink when put into acid and blue when put into alkaline. Suppose Mary, who only knows of the metaphor "litmus test", comes across a pad of litmus paper. For whatever reason, she sticks pieces in several fluids and finds that they turn pink in some liquids and just soggy in others. Upon further testing, she finds that a sample of a fluid that once turned a piece of litmus paper pink will turn other pieces pink. She then finds that if one puts a litmus paper into a fluid where it just becomes soggy, other pieces will also just become soggy in that fluid. Finally, she discovers that if one sample of a liquid turns litmus paper pink (say lemon juice), then another sample of the same liquid will also turn it pink. Even if Mary has discovered no systematic differences between the liquids which turn litmus paper pink from those that make it soggy, she is entitled to conclude that there are differences in the liquids. She is further entitled to hold that this stuff labelled "litmus paper" detects these differences with some reliability. Mary is relying on Elk Theories of the most minimal type — that one sample of milk is more like another sample of milk than either are like lemon juice, that one piece of litmus paper is pretty much like another, etc.

For establishing the reliability of a more mechanical instrument, one fulfilling more normal standards for being a Black Box, consider Bill who finds a machine labelled "Geiger counter". Although Bill has never heard the term before, he naturally enough infers that it is to count things called "geigers". But other than believing that one can count them (i.e., that "geiger" is a count, not a mass, term) Bill has no idea of what geigers are. His Elk theory is of the most minimal sort: there is a box connected to a tube and it emits clicks. In his lamentable state of ignorance, Bill certainly cannot tell whether what the rest of us know as "background radiation" is an artifact generated by the Geiger counter. Bill certainly has no independent way of determining the counter's reliability; he cannot tell whether it is missing the vast majority of geigers. Still, Bill's ignorance does not prevent him from being reasonably confident that the Geiger counter works and that it is even reliable in certain circumstances.

Suppose that whenever Bill takes his counter near a certain rock, the machine starts clicking like mad and it slows dramatically as he takes it away. After some Millean systematic trials with this rock and other rocks, moving the rock to different locations, with other Geiger counters, etc., Bill reasonably can conclude:

> Whatever my instrument detects (apparently something called a "geiger") that rock contains or emits far more of them than the other rocks around here. That would explain why it starts clicking like mad whenever I put the Geiger counter tube near it and why it slows down when I pull the tube away. This is true not only for my Geiger counter but for other Geiger counters as well. Therefore, this machine is functioning. I don't know what a "geiger" is. I know only the bare minimum about how this "counter" works. But it is detecting something and

therefore it is working with some reliability. Of course, it might also be producing artifacts — perhaps that seemingly random but on-going "background" of clicks is an example. But at least in regard to this rock it is working, although perhaps not optimally. There might be some spurious clicks and maybe many geigers are missed. Still, it is reasonably reliably detecting *something*, although I don't have any idea of what it is.

Mary's and Bill's cases are rather extreme — normally scientists have a far better idea of how a new instrument works or what it measures than Mary or Bill have. Still, they are entitled to conclude via their minimal Elk Theories that the litmus paper and Geiger counters are at least somewhat reliable in detecting something — even though they have only the vaguest idea of what this something is. And this in the absence of a substantial theory of the instrument, any theory of what is being measured, and any independent measure of the accuracy of its measurements.

One could argue for the reliability of Galileo's telescope in a similar way. In fact, we will see below that at one point Galileo essentially did this. But first, we must review the theoretical and practical difficulties facing Galileo in establishing that the telescope was reliable in the super-lunar realm.

GALILEO'S PROBLEMS IN ESTABLISHING THE TELESCOPE'S SUPER-LUNAR RELIABILITY

To see the difficulties facing Galileo in establishing the existence of the moons of Jupiter, we must first examine the difficulties he had in establishing that the telescope was a reliable source of information about the heavens. After all, he had evidence for the existence of the Jovian moons only insofar as he was justified in believing his telescope was reliable. The basic story is well-known — philosophers and historians of science have devoted a fair amount of effort to explaining/justifying Galileo's confidence in the super-lunar reliability of the telescope.[10]

Galileo started his telescopic observations of the heavens in late 1609 and soon turned his attention to Jupiter. His *Starry Messenger,* reporting his lunar, Jovian and other discoveries, was published on March 12, 1610 and it was a best-seller by 17th century standards (the 550 copies of the first printing were all sold in a week).[11] Controversy as to the reliability of the telescope soon started with a barrage of criticism from more-or-less knowledgeable critics. On April 27, Martin Horky, an assistant to Galileo's long-time rival Magini, with malicious glee, reported to Kepler:

> Galileo Galilei, the mathematician of Padua, came to us in Bologna and he brought with him that spyglass through which he sees four fictitious planets. On the twenty-fourth and twenty-fifth of April, I never slept, day and night, but tested that instrument of Galileo's in innumerable ways, in these lower [earthly] as well as the higher [realms]. On Earth it works miracles; in the heavens it de-

ceives, for other fixed stars appear double. Thus, the following evening, I observed with Galileo's spyglass the little star that is seen above the middle one of the three in the tail of the Great Bear, and I saw four very small stars nearby, just as Galileo observed about Jupiter. I have as witnesses most excellent men and most noble doctors, Antonio Roffeni, the most learned mathematician of the University of Bologna, and many others, who with me in a house observed the heavens on the same night of 25 April, with Galileo himself present. But all acknowledged that the instrument deceived. And Galileo became silent, and on the twenty-sixth, a Monday, dejected, he took his leave from Mr. Magini very early in the morning. And he gave no thanks for the favours and the many thoughts, because, full of himself, he hawked a fable. Mr. Magini provided Galileo with distinguished company, both splendid and delightful. Thus the wretched Galileo left Bologna with his spyglass on the twenty-sixth.[12]

Even agreeing with Van Helden that Horky was "ambitious, unscrupulous, and clearly very jealous of Galileo's success," Galileo was clearly facing difficulties at that time.

That spring, Magini also wrote a letter claiming that "Galileo's instrument was a deception; since colored glasses showed three suns in an eclipse, doubtless Galileo's glasses had similarly deceived him."[13] Initially, probably because of their weak telescopes, the astronomers at Rome did not see Jupiter's satellites. Horky's June 1610 pamphlet *A Very Short Excursion Against the Starry Messenger* certainly did not help. Its general tenor can be inferred by Geymonat's description of it as a "celebrated libel" and by both Kepler and Magini breaking all relations with Horky and its provoking Kepler to respond in a pamphlet *Confutation of Four Problems Proposed by Martin Horky against the Starry Messenger.*[14]

Still, when in late May 1610, Jupiter disappeared behind the Sun for a couple of months, it appears that there was no independent verification of the moons of Jupiter. This is not to say that no one else had seen them. In fact, among those who had seen them under Galileo's personal tutelage was the most important person — his prospective new employer The Grand Duke Cosimo II. Still, it "appears that only those to whom [Galileo] had personally shown the moons had seen them at all. Independent verification from outside the Galileo camp was still not forthcoming."[15] Nor were many of these criticisms unwarranted. Galileo's telescopes were not of nearly the power or quality of a good set of binoculars.

However, not everything was bleak. By April 19 of that spring, Kepler had written to Galileo a letter of fulsome praise about his telescopic discoveries. Although His Sacred Imperial Majesty's Mathematician did not have an adequate telescope to see the Jovian moons himself and acknowledged in some detail the difficulties in the theory of the telescope, he unequivocally affirmed their existence. In May 3, 1610, Kepler privately published this letter, along with a reply to various objections, under the title of *Conversation with the Siderial Messenger.* Kepler's public endorsement was a considerable relief to Galileo, who wrote on August 19 that "I thank you because you were the first one, and practically the only one ... to have complete faith in my assertions."[16] This prolonged failure of others to accept his claims could not have been a happy fact for Galileo to report. But while Kepler was convinced, many of Kepler's friends and colleagues were not. In fact, Galileo's August 19th letter was in response to Kepler's letter of August 9 which obliquely inquired about acquiring a better telescope or, at least, a list of people other than Galileo who had seen the moons. Kepler was "[w]ishing to be better able to answer the many who now opposed his support of Galileo ..."[17]

In the fall of 1610, things got better for the telescope's reputation (and Galileo's) — Jupiter's moons were observed by Thomas Harriot and friends in London as well as by a group in Aix-en-Provence. In September Kepler confirmed the discoveries.[18] By December the ageing Clavius informed Galileo that the Roman Jesuits had constructed a telescope and duplicated Galileo's observations regarding Jupiter's moons and Saturn's shape.[19] Galileo was quite pleased; at an absolute minimum, these observations showed that his claimed observations were not fraudulent or a product of his overly-fevered imagination.

Still, the criticisms did not cease. In the spring of 1611 Francesco Sizzi published *Dianomia Astronomia, Optica, Physica*, alleging that the telescope produced spurious and fictitious data.[20] Sizzi's allegation was not just theological or scholastic obscurantism, although these undoubtedly played a part.

Meanwhile things progressed. In April 1611, at the request of Cardinal Bellarmine, the four mathematicians at the Jesuits' Collegium Romanum signed a formal report confirming the existence of the satellites. On April 1, Galileo was able to report to the Grand Duke at Florence:

> I am also working to find the periods of revolutions, but they [astronomers at Rome] agree with the imperial mathematician [Kepler] in judging this a difficult and almost impossible task. Yet I have great hopes _and perhaps on my return I shall have reduced this truly atlantic labour of mine to the point of being able to predict the places and arrangements of these new planets so as to have them at every future and also each past time.[21]

By late spring 1611, the consensus on the telescope's heavenly reliability was arriving. For example, during this spring, the banker and man of letters Mark Wesler wrote a friend that although he, along with many of his astronomer friends, had initially doubted Galileo's telescopic discoveries, he was now converted as were the astronomers he knew. Clavius' acceptance had a particular impact on him.[22] Here is Van Helden's presentation of the situation in April 1611:

> Galileo's visit to Rome marked the end of the arguments about the new instrument and the reality of the phenomena it revealed. Although some persisted in their refusal to accept the new instrument, they quickly became isolated. The telescope's validity as an instrument had been amply demonstrated. Henceforth the arguments revolved around the interpretation of the phenomena it had revealed. [Van Helden in Galileo 1610/1989 p. 113]

In a letter to his friend Piero Dini written on the 21st of May 1611, Galileo presented an argument for the reliability of the telescope in reply to Sizzi and others of that ilk. This argument directly establishes the telescope's super-lunar reliability without extrapolating from the sub-lunar to the super-lunar realm, without assuming the existence of the Jovian moons and without relying on a theory of the telescope's optics. Instead, with typically Galilean panache, he argues that the results of the telescopic observations themselves guarantee that the telescope is reliable. I will discuss this argument below. By February 1612 his calculations of the satellites' periods of revolution "agreed very well with observations" and he not long thereafter published the periods of the Jovian moons. His figures are remarkably close to modern figures — within 0.05% for the modern mean synodic periods.[23] In March 1612, he first realized that his occasional failure to see a moon was due to its being in Jupiter's shadow.[24] By then, so many practiced observers had seen the moons that the serious controversy about the telescope's super-lunar reliability and the existence of the Jovian moons was over.

THE DEBATE OVER THE TELESCOPE'S SUPER-LUNAR RELIABILITY

In one sense, the debate over the reality of Jupiter's moons was primarily due to the initial absence of good telescopes. By May 1611, there were enough good telescopes and they ended the scholarly debate. How did having good telescopes resolve the central issue of artifacts? Both Horky and Sizzi had, with Galileo's guidance, seen the moons through one of Galileo's telescopes. Here is Horky's report: "On 25 April [1610] ... I saw with Galileo's telescope all four very tiny spots leaping from Jupiter, on the discovery of which he prides himself."[25] In the preface to his *Dianoia,* Sizzi reports that:

> I also, on one occasion, took part in the observations, and I saw the same phenomenon as all others present, but in some particulars it did not appear to me as to them (and I shall say more of this in my book).[26]

But the similarity of the observation reports did not show that the telescope itself was reliable in the super-lunary realm. After all, mere agreement among competent observers is still quite compatible with the Moon Illusion being an illusion — the full moon does not really shrink as it rises above the horizon. There were many reports of telescopic observations which all agreed were illusory. The first telescopes were far from achromatic, so coloured haloes around the objects were not unusual. Further, Sizzi and others argued that the telescope itself produced other illusions as well. Galileo implicitly agreed that many observations were artifacts — after all he did not proclaim that the Sun sometimes split into thirds or that the planets were surrounded by coloured haloes.

Nor were inadequate telescopes and known artifacts the only difficulties facing Galileo's claim of the super-lunar reliability of the telescope; there were well-known theoretical reasons why one could not automatically extrapolate from the the telescope's terrestrial, to its heavenly, excellence. Aristotelian physics had held that the celestial realm was made of a fifth substance which had quite different properties (e.g., crystalline, incorruptible) from the materials here on Earth. Yet, the telescope relied on the optical properties of the material which transmitted the requisite light. How then could one establish that the telescope was reliable in the heavens without begging the question as to the nature of celestial physics?[27] I will briefly review one reply available to Galileo.

In one key respect, the differences between Aristotelian physics and Galilean physics were not as substantial as is sometimes claimed. While Aristotelian physics certainly held that the laws of the super-lunar realm differed in many ways from those of the terrestrial realm, it held that some laws obtained in both. In particular, that light travelled in a straight line (barring atmospheric distortions, etc.) was as crucial for Aristotelian/ Ptolemaic astronomy as it was for Galilean/ Copernican astronomy. Without this assumption, naked-eye astronomers could not determine planetary positions, much less orbits. However, even with this common assumption, there were considerable differences between the two theories and it was an open question whether the telescope, so clearly reliable here on Earth, was also reliable in the heavens.

Establishing the telescope's terrestrial reliability is straightforward. On Earth, one can examine an object up-close and determine what it *really* looks like. Then, retreating some distance, one can see whether what one sees through the telescope closely resembles what one saw close up. Even the nasty Horky acknowledged that "On Earth it works miracles." But, because of technical transportation difficulties, this method did not work in the celestial realm; neither Galileo nor his critics examined Jupiter from up close to see if it really had moons.[28]

For a method that did work, we must turn to a fact which was very well known but rarely discussed at that time or since — the consistency of almost all naked eye celestial observations with telescopic observations. To show that this consistency played a major role in the acceptance of the telescope, we must go beyond the explicit textual arguments. This is what one would expect, if the consistency was so obvious to all that it set the limits to the very structure of the debate. To assess the evidential strength of this consistency, we must consider some possible observations that, as it happened, were not made: What if the telescopic observations by-and-large had been inconsistent with naked eye observations? What if, when Galileo had turned the telescope to the heavens, he had seen only blackness — or a solid bright green? What if through the telescope Galileo could see stars but not planets? What if he could see planets but not stars? What if Galileo could see some star-like objects through the telescope, but they did not correspond to the stars seen by naked eye observers? Or what if the moon was not visible at all, and one could see the stars it was blocking? Or ...

The considerable consistency of telescopic with naked-eye observations was crucial in establishing the telescope's super-lunar reliability. This consistency was taken for granted even by those who claimed that sometimes the telescope produced illusions.[29] But, since the telescope reported many things that naked-eye observations did not, there still was the problem of the telescope's reliability where naked eye observations were not available as a check (e.g., Jupiter's moons) as well as inconsistencies between naked eye and telescopic reports (e.g., coloured rings around bright objects).

This leads us to the second, perhaps the most natural, way of establishing the telescope's super-lunar reliability: extrapolate from its undoubted sub-lunary reliability. But such an extrapolation would beg the question against those Aristotelians who had challenged the telescope's super-lunary reliability. As we shall see, Galileo's argument does not rely on such extrapolation. Yet, although not explicitly acknowledged, the sub-lunar reliability of the telescope also must have played a substantial role in establishing the telescope's super-lunar reliability. To see this, again we must go beyond what is explicit in the texts and enquire as best we can about what would have happened if the telescope had proven to be unreliable on earth. After all, it was not a foregone conclusion that it would be reliable. Prior to hearing of Galileo's telescope, Kepler had considered building one but had decided that it probably would not work. As he wrote in his *Conversation with the Siderial Messenger*:

> After I began to work on my "Optics", the Emperor questioned me quite frequently about Della Porta's aforementioned devices. I must confess that I disparaged them most vigorously, and no wonder, for he obviously mixes up the incredible with the probable ...
>
> Furthermore, I believed that the air is dense and blue in colour, so that the minute parts of visible things at a distance are obscured and distorted. Since this proposition is intrinsically certain, it was vain, I understood, to hope that a lens would remove this substance of the intervening air from visible things. Also with regard to the celestial essence, I surmised some such property as could prevent us, supposing that we enormously magnified the body of the moon to immense proportions, from being able to differentiate its tiny particles in their purity from the lowest celestial matter.[30]

Had Kepler's surmises been correct, the telescope would not have been reliable terrestrially. There are other ways this could have happened. For example, the laws of optics might have been slightly different so that lenses might not have worked together, resulting only in a blur. Or the lenses might have produced mainly colour and shape illusions. Or they might have seriously distorted the viewed objects. That telescopes worked so well terrestrially played a major role in the final acceptance of the telescope's super-lunar reliability. This is not to say that under certain circumstances, a good argument for the celestial reliability of the telescope could not have been made even had the telescope proven unreliable in terrestrial realms. It is just to say that, in fact, the telescope's terrestrial reliability made its celestial reliability far more plausible, even though an automatic extrapolation to it would have begged the question against the Aristotelians. Had the telescope failed dismally here on Earth, it is unclear how it could ever have been relied on for astronomy.

The third way of establishing a new instrument's reliability is to build it using a well-established scientific theory. Galileo did write in the *Siderius Nuncius* that, after he had become convinced that a telescope had been invented, he was guided by theoretical considerations:

> This finally caused me to apply myself totally to investigating the principles and figuring out the means by which I might arrive at the invention of a similar instrument, which I achieved shortly afterward on the basis of the science of refraction.[31]

Despite this statement and a fuller account of how he built the telescope in his 1623 *The Assayer,* Galileo does not argue on the basis of some well-established optical theory that the telescope must be reliable. It is just as well that he did not. Van Helden comments on Galileo's sentence just quoted:

> As a professor of mathematical subjects, Galileo was thoroughly grounded in the optical theory of his day. This theory could not, however, give him much guidance in duplicating the invention.[32]

Drake summarizes the state of affairs this way:

> There was at the time no "theory of refraction" in the modern sense; all that Galileo meant by that phrase was consideration of the focal lengths of the two lenses, that is, the radii of their spherical surfaces. The power of a simple telescope is essentially a matter of the ratio of the two focal lengths, a fact that would not have taken Galileo longer to discover once he had confirmed the effect. The principal technical problem was to grind a deeper concave eyepiece than customarily used for spectacles to correct myopia; for this, a small ball could serve as a grinding tool.[33]

Feyerabend makes roughly the same point in his more polemically exuberant fashion:

> What I meant was that [Galileo] was ignorant of those parts of optics which *at the time in question* were *necessary for building the telescope*, assuming the telescope was built as a result of an insight into the basic principles of optics. What were these principles?[34]

Not only did Kepler raise objections against Galilean optics, but his own optical theory faced serious difficulties. In fact,

> ... the problem of supplying the telescope with an adequate theory that certified the existence of the phenomena seen through it continued to exercise the scientific community for the rest of the seventeenth century.[35]

Early 17th century optical theory simply was not well enough developed to justify a belief in the reliability in the telescope, much less its super-lunar reliability.

But why not argue for the telescope's super-lunar reliability in terms of ARCHED Hypotheses — that the telescopic observational reports are best explained by assuming both that the telescope is reliable and that there are moons circling Jupiter? As we saw above, Franklin has proposed such an account:

> Even granting the observation of specks of light, how do we assert their real existence as moons of Jupiter, not artifacts created by the telescope? I suggest here that it is the observed phenomena themselves that argue for their validity. Although one might imagine that a telescope could create specks of light, it hardly seems possible that it would create them in such a way that they would appear to be a small planetary system with eclipses and other consistent motions. It is even more preposterous to believe that they would satisfy Kepler's Third Law (R^3/T^2 = Constant), although that argument would not have been available until publication of Kepler's *Harmonices Mundi* in 1619, and perhaps not until later in the seventeenth century, when it was generally accepted as a law.[36]

Much of Franklin's history here is anachronistic. The scholarly debate over the telescope's super-lunar reliability had pretty much died by the summer of 1611, yet Galileo only discovered the satellite eclipses in March 1612 and the scientific debate was over well before Kepler announced his Third Law in 1619.[37] Still, Franklin's methodological point of the legitimacy of Galileo using ARCHED Hypotheses is important. On an ARCHED Hypotheses analysis, Galileo's problematic claim to have discovered moons circling Jupiter with his problematic telescope is justified because the best explanation of the observed optical regularities in the eye-piece is that both the telescope was being reliable and moons do circle Jupiter. Neither chancy hypothesis by itself can explain the optical regularities seen through the telescope. If the telescope is reliable, but there are no Jovian moons, then one would probably not have the optical regularities. Conversely, if the Jovian moons are real but the telescope is unreliable, one would probably not observe the optical regularities. Both chancy hypotheses conjoined are required to explain the regularities. Like an arch, neither hypothesis stands well supported by itself. Each requires the other for support.

Probably, the acceptance of the telescope as well of as the Jovian moons was accelerated by the fact that positing the reality of the Jovian moons so naturally explained the telescopic observations. Using ARCHED Hypotheses this way is valid even though the initial proposals were primarily qualitative—when he wrote the *Siderius*, Galileo had determined the period of only the most distant moon and could only report that the inner moons are faster than the outer.[38] Although ARCHED Hypotheses may well have played a major part in the acceptance of the telescope's reliability, they are not the full story. Galileo's letter to Dini provides an additional part.

GALILEO'S USE OF ELK THEORIES TO SHOW THE REALITY OF THE JOVIAN MOONS

Although Galileo no doubt implicitly relied upon the ARCHED argument and the argument from the consistency of telescopic and naked-eye observations, in his May 1611 letter to Piero Dini, Galileo produced a different argument. This neglected argument is what I am interested in here, for it relies on Elk Theories and the Jovian telescopic "observations" to establish that the observations are reliable; unlike the ARCHED Hypotheses argument, it is quite independent of the Jovian moons hypothesis.

Galileo first proclaims that he has not been deceived and is not a fraud. Then, perhaps anticipating "The more he proclaimed his honesty, the faster we counted the silver," he soon drops the honesty theme and continues:

> ... it is not only my telescope, or those that I built, which show the four Jovian Planets, but all other telescopes as well, no matter where they were built or by whom, provided that they have been properly made and show objects in a large and distinct way. With all these instruments, no matter where they are used, it is possible to see the same changes from night to night and the same arrangements of these Planets. Therefore, those who would want to maintain that these phenomena are illusory, will find it very difficult to explain why all instruments, both big and small, both long and short, make just the same errors and to explain why, among the innumerable visible objects, all the illusions appear around only the star of Jupiter. Further, I add that if anyone is confident that it is possible to construct a telescope that would, by means of an illusion, make non-existent multiple lights or other things appear around a star or light or any other particular object, and that such an appearance would happen around only one object and not around the others, please have him build such an instrument, for I promise to pay 10,000 scudi for it. And if my telescope had the ability to make something appear other than it really is, I would not exchange it for any treasure you can name. And this is enough to unmask the belief that there is an illusion, a belief which would be eliminated from anyone by one single glimpse through the instrument.[39]

Galileo here reports two striking phenomena: independent observers in different locations using well-made telescopes report seeing (1) the same complex, changing pattern of lights (2) near Jupiter and only near Jupiter. If early Tuesday evening someone in Rome reports three lights in a straight line to the right of Jupiter, independent observers in Padua report the same thing. If, two days later, Galileo reports three lights to the left of Jupiter, independent observers in Rome report the same thing. Further, these observers do not see such lights around Mars, Venus, Sirius or Polaris. These phenomena are at the heart of Galileo's argument.

Notice, that although he clearly was confident of his Jovian moon explanation for the pattern, Galileo is not here using it as an ARCHED Hypothesis. That is, theoretically one could accept this argument that the telescope is not producing an artifact and yet reject the claim that there are moons circling Jupiter. His argument is structurally similar to our earlier imaginary examples: Mary's, that litmus paper was reliably detecting something in some liquids that turns the paper pink, although she had no idea what it was; and Bill's, that the Geiger counter was counting geigers, whatever they are. While Galileo does not spell out his argument in detail, it is not hard to reconstruct some of his reasoning.

First, we should briefly note that Galileo's criterion for a "well constructed" telescope is not circular — a well constructed telescope is one that makes objects "visible in a large and distinct way." This was a serious practical issue, for as he wrote on March 19, 1610, "... spyglasses that are most exquisite and capable of showing all the observations are very rare, and among the sixty I have made, at great cost and effort, I have been able to find only a very small number."[40] Identifying those very few telescopes can be done on Earth and is totally independent of whether one is able to see the Jovian moons through them. Then, since the patterns in reported observations were reported by many independent observers, using independently made telescopes, the patterns are not explicable by fraud or by Galileo's self-deception.

Commendably putting his 10,000 scudi where his mouth was, Galileo claims that no one could build a telescope which could generate "lights or other things" around an object (presumably celestial, but not necessarily) and not around any other objects. In particular, there could not be a telescope that could generate such patterns of lights around some object other than Jupiter and not generate one around Jupiter. Galileo is claiming that it is impossible for any telescope to create such patterns as an artifact. This claim is, I am proposing, based on an Elk Theory of the telescope. As a theory, the Elk Theory is quite weak — much weaker, in fact, than the "baby optics" that Feyerabend rightly derided.[41] It is a theory that couldn't begin to tell how to build a telescope, couldn't even predict that the telescope would work, and certainly couldn't explain how it worked; still it has enough content to enable us to conclude that certain things are not artifacts. What then was this very weak but important early 17th century Elk Theory of the telescope?

Since Elk Theories are so obvious that generally no one bothers recording them, we must indulge in a little rational reconstruction here. Telescopes were not Black Boxes — whatever the quality of early 17th century optical theory, Galileo and his critics knew certain things. They knew that telescopes consisted of two lenses at opposite ends of a tube and that the lenses remain stationary relative to each other. They knew that the telescope had no internal moving parts. While lacking anything approaching an adequate theory of optics, still there were Elk Theories of lenses. First, the lenses did not generate light. Second, they were rigid physical objects. Third, whatever the laws under which they worked, under similar conditions they did similar things. For example, if a lens enlarged an object's image at a certain distance, it always enlarged objects at roughly that distance and the enlargement was of roughly the same amount.

Given a constant object at a constant distance, a lens reliably produced a constant image. These facts are clearly ascertainable without our having anything worthy of being called "a theory of telescopic optics". But, however weak, this Elk Theory of the telescope sufficed to ensure that the patterns of Jovian observations were not artifacts. With fraud and self-deception dismissed, Galileo was left with two candidate hypotheses: (1) the patterns are artifacts produced by the inner-workings of the telescope, perhaps in interaction with the observer's eye and (2) a well-made telescope reliably transmits the information that there are some things in Jupiter's direction when viewed from the Earth. The telescope's Elk Theory shows that the artifact hypothesis is extremely implausible.

To be plausible, the artifact hypothesis must at least be *consistent* with two aspects of the data: the lights are only seen near Jupiter regardless of where Jupiter is and they change in regular ways. If the lights are artifacts, somehow these artifacts must be generated only when the telescope is aimed at Jupiter, independent of whether Jupiter is near the horizon or high in the sky, near the Moon or Sun or far from the Moon or Sun. Moreover, the artifact hypothesis must acknowledge that Jupiter's naked eye appearance resembles that of the brighter stars and the other planets. Further, Jupiter's brightness varies during the course of the year — so the artifactual explanation of the lights cannot depend on any precise detail of Jupiter's light. Also, the lights are not a function of Jupiter's location on the celestial equator — the other planets travel near the celestial equator and yet do not have similar lights seen near them. To be plausible, the artifact hypothesis must also be consistent with the changing patterns of lights near Jupiter and only near Jupiter.

I am not so concerned here about the long-range prediction of the pattern, as about the simpler fact that the purported artifacts systematically differed over time. For each time, the lights seen in differently-located, differently-manufactured (big and small, long and short) telescopes were the same. The lights were always arrayed on a straight line that intersected Jupiter. But what was seen in well-made telescopes at one hour was different from what was seen a few hours later, much less the next day. Further, while precise prediction of the lights' positions relative to Jupiter proved difficult, the patterns did not change randomly. That is, they changed in a slow continuous way. Finally, the artifact hypothesis must be consistent with the fact that although Jupiter travels near the ecliptic in front of seemingly randomly placed fixed stars, the pattern is distinctly non-random and thus apparently independent of the background of fixed stars.

By themselves, these facts are not inconsistent with the artifact hypothesis. With modern electronics, we could create a machine to distinguish Jupiter's light and then create whatever artifacts we want and array them as we please. The difficulty comes in making the artifact hypothesis consistent with the telescope's Elk Theory. The Elk Theory tells us that the telescope is a simple object, consisting of stationary parts and that given similar inputs lenses give similar outputs. How could a telescope of unchanging parts reliably distinguish Jupiter from optically similar objects such as Saturn or Sirius? If Jupiter were visually quite different from other celestial bodies (as the Sun and the Moon are) then a telescope's producing artifacts about it would be more intelligible. But in Jupiter's case, how could a telescope produce artifacts near it and not near optically similar items?

Further, how could a telescope of unmovable parts produce slowly modifying artifacts — ones sometimes to the left of Jupiter and ones sometimes to the right? It could not be the telescope's immediate environment, because the same reports came from Rome and Florence. It was seemingly independent of the angle of the telescope, because the pattern was seen both when Jupiter was near the horizon and high in the sky. This phenomenon was unlike the Moon Illusion which happens whenever the Moon is near the horizon. It is unlike the Moon Illusion in another way — the Moon always looks larger as it starts to rise. But the pattern of lights near Jupiter is not constant — the location and the numbers regularly varies over time. From such considerations, Galileo concluded that the telescope is not producing an illusion, but is, at least in this controversial instance, reliably reporting what is in the heavens.

I suspect that one advantage for Galileo of using Elk Theories comes from the epistemic situation that Galileo found himself in. Sizzi and others had proposed that the telescope itself produced illusions and there was no well-supported optical theory available to him. Galileo's argument is essentially that *this* sort of regularity, regardless of the cause of it, is not like the sort of illusions that Sizzi describes (e.g., the three Suns in an eclipse). Those illusions (and Galileo agrees that they are illusions) would be produced by any suitably bright object and would not change over time. But the bright lights are not seen near Mars and the pattern of changes seems to be quite independent of who looks at Jupiter, from where, and under what conditions. Since the telescope has no moving or changing parts, it is very difficult to imagine how it could produce dots that changed so regularly around just that one object.

More generally, Galileo's argument rests on taking (more or less) seriously the theory that the telescope is unreliable and proposing that there are phenomena which appear to be clearly inconsistent with that theory. In effect, Galileo views the telescope's super-lunar reliability as a hypothesis, to be treated like any other hypothesis. He concludes, reasonably enough, that the "telescope-is-producing-an-artifact-near-Jupiter" hypothesis seems to be inconsistent with his Elk Theory of the telescope. But the Elk Theory (the telescope's fixed parts can not give rise to regular changing patterns of lights) has no controversial elements of optical theory and so every one can agree with it.

Of course, like any other argument for an instrument's reliability, such an argument is not indubitable.[42] This Galileo implicitly acknowledged a year later.

Several months after discovering the Jovian moons, Galileo had turned his telescope on Saturn and observed what we now know to be its rings. He initially interpreted them as two objects which permanently remained near Saturn, as he reports in his polemic against "Apelles":

> ... I have resolved not to put anything around Saturn except what I have already observed and revealed — that is, two small stars which touch it, one to the east and one to the west, in which no alteration has ever yet been seen to take place and in which none is to be expected in the future, barring some very strange event remote from every other motion known to or even imagined by us. ... [Contrary to Appelles' claim] I, who have observed it a thousand times at different periods with an excellent instrument, can assure you that no change whatever is to be seen in it. And reason, based upon our experiences of all other stellar motions, renders us certain that none ever will be seen, for if these stars had any motion similar to those of other stars, they would long since have been separated from or conjoined with the body of Saturn, even if that movement were a thousand times slower than that of any other star which goes wandering through the heavens.[43]

Given Galileo's categorical proclamation that his telescope and observations were superior to "Apelles' " and that Saturn's companions had no separate motion, it was disconcerting later to have to report that he could no longer see Saturn's companions. Here is his letter of December 1, 1612:

> ... let me now return freely and without interruption to the studies which I shall shortly publish together with the consideration of other celestial novelties. Among these is an unexpected wonder regarding Saturn which has recently come to disturb me, of which I must render an account to Your Excellency.
>
> About three years ago I wrote that to my great surprise I had discovered Saturn to be three-bodied; that is, it was an aggregate of three stars arranged in a straight line parallel to the ecliptic, the central star being much larger than the others. I believed them to be mutually motionless, for when I first saw them they seemed almost to touch, and they remained so for almost two years without the least change. It was reasonable to believe them to be fixed with respect to each other, since a single second of arc (a movement incomparably smaller than any other in even the largest orbs) would have become sensible in that time, either by separating or by completely uniting these stars. Hence I stopped observing Saturn for more than two years. But in the past few days I returned to it and found it to be solitary, without its customary supporting stars, and as perfectly round and sharply bounded as Jupiter. Now, what can be said of this strange metamorphosis? That the two lesser stars have been consumed, in the manner of the sunspots? Has Saturn devoured his children? *Or was it indeed an illusion and a fraud with which the lenses of my telescope deceived me for so long—and not only me, but many others who have observed it with me? Perhaps the day has arrived when languishing hope may be revived in those who, led by the most profound reflections, once plumbed the fallacies of all my new observations and found them to be incapable of existing!*
>
> I need not say anything definite upon so strange and unexpected an event; it is too recent, too unparalleled, and I am restrained by my own inadequacy and the

> fear of error. But for once I shall risk a little temerity; may this be pardoned by Your Excellency since I confess it to be rash, and protest that I mean not to register anything here as a prediction, but only as a probable conclusion. I say then, that I believe that after the winter solstice of 1614 they may once more be observed...[44]

Fortunately for Galileo, the "companions" were seen again and the value of the telescope was not again threatened by the "profound reflections" of his critics.

University of Melbourne, Parkville, Victoria, Australia

NOTES

* I appreciate the help of Keith Hutchison, Ross Phillips, the La Trobe Philosophy Department, Alan Franklin, Deb Dowling, Rosemary Smith and Michael Ellis. I am particularly grateful for my editor's patience.

1 Although I find Feyerabend's challenge illuminating, many of the details of his historical and philosophical evidence have been so roundly and persuasively criticized, that I will not deal here with his arguments [1975]. For critics, see, for example, Machamer 1973, Frankel 1978, Brown 1985 and Thomason 1994. I will also by-pass Wallace's 1992 analysis of Galileo's argument because it does not address the problematic status of the telescope.

2 For more details of Butler's satire, see Wasserman 1967, pp. 39-42.

3 Van Helden in Galileo 1610/1989, p. 89 n.

4 Hacking 1983, p. 23.

5 Franklin 1990, pp. 103-15, esp. p. 104.

6 *Ibid.*, p. 110.

7 For more detail on Franklin's account of Millikan, see Franklin 1986, p. 169.

8 Detailed arguments both for the intuitive rationality of using ARCHED Hypotheses as support for two doubtful hypotheses and for its consistency with Bayesian probability are given in Thomason 1994.

9 Python 1989, p. 119. The fate of Miss Elk's second theory shows that even Elk theories are not infallible. She no sooner says, "My second theory states that fire brigade choirs seldom sing songs about Marcel Proust," than the fire brigade starts singing the Proust song.

10 Drake 1979 and 1983 are essential reading for anyone interested in the history of these matters.

11 Drake 1978, p. 158. (Actually, such sales would be quite a good start for a late 20th century science book!)

12 Galileo 1610/1989, pp. 92-93.

13 Drake, 1978, p. 160.

14 Geymonat 1965, p. 44, and Drake 1983, p. 147.

15 Van Helden in Galileo 1610/1989, p. 02.

16 Rosen 1965, p. 71, note 85.

17 Drake 1978, p. 162.

18 Galileo 1610/1989, p. 91.

19 *Ibid.*, p. 109.

20 Galileo 1890-1910 3:12 and 201-250 especially 3:211.

21 Drake 1979, p. 83.

22 Drake 1978, pp. 165-66.

23 Débarbat and Wilson 1989, p. 144.

24 Drake 1979, p. 89.

25 Drake 1983, p. 148.

[26] *Ibid.*, p. 152.
[27] In recent years, some philosophers have tended to make a Big Deal about these difficulties — in my view, too Big a Deal. For an insightful presentation of the difficulties facing Galileo's claims of super-lunar reliability and why they were not as serious as some philosophers have proposed, see Chalmers 1990, pp. 41-60, especially pp. 54-55.
[28] The advantages of such close-up observations were apparent, if not achievable. Consider Galileo's joke at the expense of the philosopher Giulio Libri (1550-1610). Although (or perhaps because) Libri had been Galileo's colleague at both Pisa and Padua, he refused to look through the telescope, thus inadvertently ensuring his posthumous fame. After Libri's death, Galileo joked that perhaps on his way to Heaven Libri would observe Galileo's discoveries. This joke represents a considerable advance in astronomical methodology—a 17th century version of the *Voyager* space probe. The major difference was Voyager's greater success in relaying its observations back to those of us mired in the mortal state.
[29] Yet, such consistency was not necessary. For example, if Angels and the Lord's Throne as described by Dante were seen through the telescope, this would have decisively established the telescope's super-lunar reliability. From this atheist's wildly anachronistic perspective, this would have been the most spectacular case of ARCHED Hypotheses in history.
[30] Rosen 1965, pp. 17-18.
[31] Galileo 1610/1989, p. 37.
[32] *Ibid.*
[33] Drake 1978, p. 140.
[34] Feyerabend 1975, p. 115.
[35] Van Helden in Galileo 1610/1989, p. 89.
[36] Franklin 1986, pp. 168-69. Franklin's account comes just before his comments on Millikan's oil drop experiment cited above. Franklin later paraphrases his account of Galileo this way: "Even without such a theory, we can validate observations. Sometimes the phenomena themselves provide such evidence. The observations of the moons of Jupiter by Galileo were extremely unlikely to have been artifacts of the telescope, as suggested by Cremonini, because they appeared to represent a consistent planetary system and even obeyed Kepler's Third Law (R^3/T^2 = constant)." Franklin 1986, p. 192.
[37] This is Van Helden's summary of the situation by the spring of 1611: "The moons of Jupiter were the test case of instrument quality, and questions about their existence dominated discussions about the new discoveries. Although this part of the evaluation was over by the spring of 1611..." Galileo 1610/1989, p. 89.
[38] Galileo 1610/1989, p. 84.
[39] Galileo 1890-1910 XI, pp.106-7. Felix Siddell considerably improved this translation.
[40] Galileo 1610/1689, p. 92.
[41] Feyerabend 1975, pp. 115.
[42] See Rasmussen 1993 for an illuminating history of a long-lasting artifact that seemingly satisfied several of Franklin's proposed strategies.
[43] Drake 1957, pp. 101-102 (my italics), translation of *History and Demonstrations Concerning Sunspots and their Phenomena*, 1613.
[44] Drake 1957, pp. 143-44 (my italics).

REFERENCES

Brown, H.I. (1985) 'Galileo, the Telescope and the Eye', *Journal of the History of Ideas*, 46: 487-501.

Butler, S. (1854) *Complete Poetical Works*, Volume II.

Chalmers, A. (1990) *Science and Its Fabrication*, Open University Press, Milton Keynes.

Débarbat, S. and Wilson, C. (1989) 'The Galilean Satellites of Jupiter from Galileo to Cassinni, Romer and Bradley' in Taton and Wilson (eds.) (1989), pp. 144-57.

Drake, S. (1957) *Discoveries and Opinions of Galileo*, translated with an introduction and notes by Stillman Drake, Anchor Books, New York.
— (1976) 'Galileo's First Telescopic Observations', *Journal for the History of Astronomy*, 7: 153-68.
— (1978) *Galileo at Work.* University of Chicago Press, Chicago.
— (1979) 'Galileo and Satellite Prediction', *Journal for the History of Astronomy*, 10: 75-95.
— (1983) *Telescopes, Tides, and Tactics: A Galilean Dialogue about the Starry Messenger and Systems of the World.* University of Chicago Press, Chicago.
Feyerabend, P. (1975) *Against Method.* First Edition, Verso Press, London.
Finocchiaro, M. (1989) *The Galileo Affair: A Documentary History*, University of California Press, Los Angeles.
Frankel, H. (1978) 'The Importance of Galileo's Nontelescopic Observations Concerning the Size of the Fixed Stars', *Isis*, 69: 77-82.
Franklin, A. (1986) *The Neglect of Experiment.* Cambridge University Press, Cambridge.
— (1990) *Experiment Right or Wrong.* Cambridge University Press, Cambridge.
Galilei, Galileo (1890-1910) *Opera*, 20 volumes; National Edition edited by Antonio Favaro, Florence.
— (1610/1989) *Sidereus Nuncius or The Sidereal Messenger*, translated with commentary by Albert Van Helden, University of Chicago Press, Chicago.
Geymonat, L. (1965) *Galileo Galilei: A Biography and Inquiry into his Philosophy of Science*, translated by Stillman Drake, McGraw-Hill Book Company.
Hacking, I. (1983) *Representing and Intervening*, Cambridge University Press, Cambridge.
Kosso, P. (1989) *Observability and Observation in Physical Science*, Kluwer, Dordrecht.
Kuhn, T.S. (1957) *The Copernican Revolution*, Harvard University Press, Cambridge, Mass.
Machamer, P.K. (1973) 'Feyerabend and Galileo: The Interaction of Theories, and the Reinterpretation of Experience', *Studies in History and Philosophy of Science*, 4: 1-46.
Python, Monty (1989) *Monty Python's Flying Circus—Just the Words*, Volume Two, Methuen-Mandarin Paperbacks.
Rasmussen, N. (1993) 'Facts, Artifacts, and Mosesomes: Practicing Epistemology with the Electron Microscope', *Studies in the History and Philosophy of Science*, 24: 227-65.
Rosen, E. (1965) *Kepler's Conversation with Galileo's Sidereal Messenger*, Johnson Reprint Company.
Segre, M. (1991) *In the Wake of Galileo*, Rutgers University Press, New Brunswick, N.J.
Stubbe, H. (1670) *The Plus Ultra reduced to a Non Plus: Or, a Specimen of some* Animadversions *upon the* Plus Ultra *of Mr. Glanvill, wherein sundry* Errors *of some* Virtuosi *are discovered, the Credit of the* Aristotelians *in part* Re-advanced; *and Enquiries made about the Advantages of the* Ancient Education *in* England *above the Novel and Mechanical.* Printed for the Author.
Taton, R. and Wilson, C. (eds.) (1989) *Planetary Astronomy from the Renaissance to the Rise of Astrophysics. Part A: Tycho Brahe to Newton.* Volume 2 of *The General History of Astronomy* under the general editorship of Michael Hoskin, Cambridge University Press, Cambridge.
Thomason, N. (1994) 'The Power of ARCHED Hypotheses: Feyerabend's Galileo as a Closet Rationalist', *British Journal for the Philosophy of Science*, 45: 255-64.
Van Helden, A. (1974) ' 'Annulo Cingitur': The Solution of the Problem of Saturn', *Journal for the History of Astronomy*: 155-74.
— (1989) 'Galileo and Telescopic Astronomy' in Taton and Wilson (eds.) 1989.
Wallace, W. (1992) *Galileo's Logic of Discovery and Proof. Boston Studies in the Philosophy of Science* Volume 137, Kluwer, Dordrecht.

ADRIAN HEATHCOTE

OF CROWS AND QUARKS: REFLECTIONS ON THE LAWS OF QUANTUM MECHANICS

I

To the metaphysician the puzzle about laws of nature arises from their apparent modal nature, their ability to sustain counterfactual assertions. The various *theories* about natural laws differ in the degree to which they respect that intuition about modality and in how they try to accommodate it. At some point however, and preferably sooner rather than later, such metaphysical theories must interact with our theories about the world, our science. A metaphysics that does not dovetail with our best science is surely too much *meta* and not enough *physics*.

Yet as the Humean tradition makes clear, it is by no means obvious that this modal intuition can or should be respected even in the case of classical science. For the Humean there are only universal generalizations describing the behaviour of objects — there is no necessity, no *must-ness*, that attaches to these descriptions: metals may expand when heated but it is not *necessary* for them to do so; the next piece of metal may expand when heated — or it may not — but if it does then that is all that is needed for it to be in accord with the law (if it is one) that all metals expand when heated. Actuality, not necessity, is the mark of the Humean law of nature.

Humeans typically make some concessions here; after all not every universal generalization is a law of nature, surely, even if every law of nature is a universal generalization. What is needed is something that distinguishes natural laws from *accidental* generalizations. The problem for Humeans is to try to draw that distinction in such a way that it does not collapse into a contrast between the *accidental* and the *necessary*. The usual way out is to make the distinction a subjective matter. For A.J. Ayer laws of nature are those generalizations that we esteem, that we accord a special place in our account of nature.[1] Just why we should esteem some generalizations more than others is not clear — we can't esteem them because we regard them as laws of nature, on pain of circularity; our esteem just has to be taken as a brute fact. A better answer is given by the Ramsey-Lewis theory in which laws are theorems of the simplest and strongest set of axioms that describe the natural world. The subjective component of this Humean view is still present, in some measure at least, in the notion of simplicity but it looks less arbitrary than Ayer's account.[2]

P.J. Riggs (ed.), Natural Kinds, Laws of Nature and Scientific Methodology, 145–161.

The Humean tradition speaks in the first instance of objects that behave in certain regular ways: laws of nature are generalizations about those objects. However the Humean tradition has, for purely historical reasons, diverged from Hume himself on the question of properties. In Hume's writing we find no reluctance to countenance the existence of real properties that are instantiated in ordinary objects. In his discussion of inductive inference Hume speaks of bread nourishing, candle flames burning, etc; these are, for Hume, qualities that inhere in the objects. Indeed, in his discussion of causation, the obvious inherence of certain qualities in physical objects is set as a contrast with the case of power or necessary connection (where we have no impression and therefore no reason to believe that such a quality exists). It is not, of course, that Hume has any *theory* of universals; far from it: he is simply a common-sense realist in such matters who has found no reason to be sceptical. His scepticism about induction is not a scepticism about properties being instantiated in objects but rather whether properties that have gone together in the past — bread nourishing, flames burning, etc — will go together in the future.

So although a Humean account of laws of nature will, in the first instance, speak of generalizations over objects it can be recast to speak of properties. Thus, for all x if x is a crow then x is black, describes a regularity at the level of objects but, also, a relationship between properties: one property *goes along with* another. A law of nature links together the instantiation of one property with the instantiation or behaviour of another. For the Humean this *going along with*, this linkage between properties, is irreducible, unexplained and, most importantly, entirely accidental. Properties that have gone together in the past need not go together in the future; and the possibility that they not go together is sufficient to show that their going together was always an accident: there is no necessity, no metaphysical glue, that need keep them together. This is the point of Hume's argument for inductive scepticism: at any moment flames and heat may cease to go together so our inference from flame to heat must always be uncertain. What accident has joined chance may at any moment put asunder.

Of course though Hume is happy to countenance properties being instantiated in objects many Humeans have not gone along with him here. For many Humeans talk of properties is simply a convenient shorthand for talking of something else: predicates, classes, or resemblances. Yet a Humean who is, say, a resemblance nominalist has to think that our putative law, all crows are black, is another way of saying that those things that resemble this thing (a crow) in one respect (its genotype) will resemble it in another (a phenotype–blackness). What's more, they have to do it without making use of the properties mentioned in parentheses! It is difficult to resist the feeling that the Humean nominalist, who, after all, must try to make sense of a *relation*, however accidental, between different resemblances is making life intolerably difficult for him(her)self here.

The nominalist paraphrase is evacuated of substantial content: it is an equation with *only* variables, forever unsolvable and forever unintelligible.

To make sense of Humeanism, therefore, we should allow it to make full use of talk of properties. We have then two ways of stating the law that all crows are black: either as an explicit first-order quantification $(x)(Cx \supset Bx)$, with C and B standing for the obvious, or as a relation between properties: *crowness* goes along with *blackness*. I take it that these different ways of speaking do not differ with respect to their *Humeanism* but only perhaps with respect to their commitment to properties (I say 'perhaps' because I'm by no means convinced that they do, Quinean dogma notwithstanding). They do not differ in their Humeanism because the lack of necessity in the universal generalization is mirrored by the lack of necessity in the two properties 'going along together'. *Going along together* is a matter of accident, of the *concurrence* of two properties. As Hume says, all things 'seem entirely loose and separate'. The concurrence of two properties is a brute fact, unexplained and irreducible.

A Humeanism that traffics in properties is an advance over one that doesn't, but the basic implausibility of the position remains untouched. Whether one expresses the view as an accidental concurrence of properties or as a first-order universal generalization we confront the problem that genuine accidents cannot be *expected* to continue. If we have reason to believe that it is nothing more than an accident, a matter of fact but not necessity, that crows have black feathers then, from the standpoint of our current epistemic situation, we not only have no reason to expect that the next crow will be black we have positive reasons to believe that it will not be. If a fair coin has been tossed a thousand times and has come up heads then, *since it is fair,* the probability of it coming up heads on the next toss is still ½. That probability cannot change without our committing some form of the Gambler's Fallacy. Probability theory cannot rule out the possibility of long runs of heads — it is not up to *it* to legislate against miracles — it can only guide us given the information that we have. If we were ignorant as to the fairness of the coin then a long run of heads might lead us to form an estimate of bias. Lewis' Principal Principle is in fact a formula for making such revisions. But if we are assured that the coin is unbiased then that is just the evidence we need to keep our prediction of the coin's outcome at the next toss as ½. Not to do so will force us into a Gambler's Fallacy.[3]

Now the Humean is forced to enact this opening from Stoppard's *Rosenkrantz and Guildenstern are Dead* with all of our putative laws. Just as it is an accident if a fair coin comes up heads at every toss so it is an accident that crows are black, that a flame is hot, that metals expand when heated and so on. All we have here is an accidental concurrence of properties. Hume's inductive scepticism is an argument against the concurrence of properties being anything more than accidental, against the *necessity* of their concurrence. But being *assured* that the concurrence is accidental is the same as being assured that the coin was unbiased: the coin always happens to come up heads just as metals in fact just happen to expand when heated; the coin does not *have* to come up heads, it is not biased, and nor does the metal *have* to expand when heated. But now the inductive scepticism that ruled out necessity threatens to go too far in the opposite direction: we should now believe, just as with the coin, that the next piece of metal that is heated has only a probability of one half of expanding. Ditto with the next crow — it might be black, but it might also be white, green, or striped. And so on with all of our 'laws of nature'. Inductive scepticism, and hence the Humean conception of laws of nature, has all the vertiginous drama of a man standing on the edge of a cliff at night — one step beyond and everything is chaos, everything unknown.

(It is worth recalling here an elementary feature of probability theory that the Humean too easily forgets. Two events are independent if there is no *causal* connection between them: a fair coin has a chance of one half of coming up heads each time it is tossed — these tosses are then independent. By eliminating substantial causal connections the Humean makes *all* purportedly causal events independent — that is why Humeanism runs into such trouble with laws of nature.)

The argument that we've just given suggests that a Regularity Theory of laws of nature logically entails Humean inductive scepticism. A careful reader of Hume's texts would probably not be very surprised at this; after all Hume speaks in section IV of the *Enquiry* of causes and effects that have gone together in the past not needing to go together in the future:

> ... it implies no contradiction that the course of nature may change, and that an object, seemingly like those which we have experienced, may be attended with different or contrary effects. May I not clearly and distinctly conceive that a body, falling from the clouds, and which, in all other respects resembles snow, has yet the taste of salt or feeling of fire?[4]

However, a *very* careful reader of Hume will note that he does not, in fact, speak of *snow* tasting of salt or feeling of fire but only of something that *resembles* snow doing so. But of course something that resembles snow may feel warm to the touch without this implying that *snow* may feel anything but cold. Most readers of Hume seem to ignore this point, however, and take Hume to be saying that the course of nature may change so that *snow itself* no longer occasions feelings of cold. In this way one and the same argument does both epistemological and ontological damage, for the argument is taken to show that we have no reason to believe that future snow will feel cold, and also that there was no necessary connection linking snow and the feeling of cold. Things may fall apart at any moment and so belief that is conditioned by the past is no guide to what may happen in the future.

Thus the Regularity Theory of laws implies inductive scepticism and this is why they are so closely associated in Hume's work. If many defenders of the

Regularity Theory have not felt the force of this implication it is because a large number of Humeans seem to exist in a state of *genteel bad faith*, where the quite bizarre and counter-intuitive nature of the Regularity Theory is shielded from full view by certain tranquilizing formulations. After all, it sounds quite harmless to say that laws are unmodalized universal generalizations; it certainly sounds better than saying that properties that have gone together in the past need not, but as a matter of fact will, go together in the future. The latter smacks of miraculousness, of the charlatan magician's tawdry top hat and doves.

The essential metaphysical problem of laws of nature can be put thus: there is nothing between accidentality and necessity that will distinguish laws of nature from simple contingencies on the one hand and necessities on the other. My argument so far has suggested that the Humean view is wrong (and clearly no amount of subjective 'add-ons' will solve the problem) but gives us no idea as to what the right view might be. All we know so far is that our inductive predictions, our predictions based on our estimation of *nature's bias*, as one might call it, will be continually mistaken. It will be no more than an even bet that the sun will rise tomorrow.

We can make a start on trying to solve this problem if we look again at the Humean claim that the *going along with* in the law that *crowness* goes along with *blackness* is merely an accidental concurrence. It seems quite wrong, of course, even to the Humean, but saying why is not easy: we must look at the crows more closely.

What makes something a crow is a certain complex property, its genotype. This is a vast complex of information in the chromosomes that will result in particular phenotypic traits given a normal environment. But one part of that genotype will normally result in black feathers; so part of what makes something a crow makes it black. It is not that one property just happens to go along with another property but that one property contains the other property (or to be more precise, the underlying structure that is one property results in the other). The relation is not side-by-side accidental occurrence but proper inclusion, the relation of part to whole.

But the issue is, of course, more complex than this suggests. To judge that it is a law that *all crows are black* is effectively the same as judging that the identity conditions for crows, or the crow-genotype, include as an essential part the genes for black feathers. Nor is this judgment about essential parts implausible. The identity conditions for a whole and its parts involve us in a Sorites: at some point *some* part has got to be essential to the identity of the whole or you will be able to take away *all* parts and the thing still be itself. This is not to say of course that some *particular* part must be essential — some special grain of sand that if removed will make the heap something other. That is where the analogy of part/whole to grains in heaps breaks down. But a part may be so large or otherwise significant that subtracting it from the whole means that we no longer have the same whole that we once had. To judge that all crows are black is a law, or that, contrariwise, nothing that is non-black is a crow (or *can be* a crow) is to judge that blackness is an essential part of what makes something a crow. (Of course we must not allow ourselves to be distracted by the details of this example: the real possibility of non-black crows, albinos, shows that, in the judgment of biologists it is *not* a law that all crows are black. So much the worse for hoary old philosophical examples.)

The point is that laws of nature express a relation between properties that seems more intimate than the Humean's accidental concurrence. But for this view to work it must be able to be applied to examples of genuine laws of nature — not ersatz ones like all crows are black. Presumably the solution for real laws will be very different from the part-whole solution that seems to work for the ersatz laws. Thus we seem to have arrived, as a first approximation, at the view that was developed by Fred Dretske, Michael Tooley and David Armstrong, that laws are relations (of more than just accidental concurrence) between properties (universals). Or at the very least we have an argument that seems to lead to the rejection of Humeanism for reasons similar to those expressed by those authors.

Yet there are differences too. Like Armstrong, for example, I think that the Regularity Theory implies inductive scepticism, but where Armstrong thinks that laws are our *best explanation* for the success of inductive inferences and that probability alone will not do, I think that probabilistic inferences are enough.[5] On my view, since the Regularity Theory implies inductive scepticism, it must be the case that *if* probabilistic inference shows inductive scepticism to be false that it also show the Regularity Theory to be false. But something weaker than a Regularity account of laws may also imply inductive scepticism and so showing that inductive scepticism is untenable does not show that everything is law-like: thinking that there are mere regularities is only one (extreme) way of misrepresenting nature's bias.

The problem that I see with Armstrong's view is that there are many situations in which inductive scepticism seems false that do not appear to be underwritten by laws of nature. For example, if I have a slightly biased coin and I do not know to which side it is biased then, if on ten trials heads turns up eight times it seems to be a perfectly good inductive inference to think that heads is more likely to turn up on the next trial, because the coin is more likely to be biased toward heads. We don't reach for a law of nature in this case because it does not seem to be required: the bias of the coin explains all. On Armstrong's view there are either going to be far too many laws of nature or far too few successful inductive inferences.

Nevertheless, I agree with Armstrong that our prediction that the next piece of metal will expand when heated is bound to be correct because *in this case* there is a law of nature explaining the individual instances. Laws of nature show nature at its most biased and therefore inductive scepticism at its most jejune. But *any* bias in nature will show-up the problems in inductive scepticism. Laws of nature are indeed our *best explanations* for the extremal cases of probabilistic inferential success but it does not follow that there will be no success where laws are not there to do the explaining. Inductive scepticism is false because nature is often both biased and constant in its bias, and the laws of nature are a special case of this (and in fact the one that Hume focussed on). The more we understand of nature's bias the better our inductive predictions are. Thus we know that the unobserved will resemble the observed — in the fallible way we know anything — because we have every reason to believe that our sampling was random. Humeanism is false because in getting rid of the *connections* between things it made the concurrence of properties an unexplained miracle and this leads to inductive scepticism. But since inductive scepticism is false (and even Hume, we must remember, found it an unbelievable consequence!) so must be the Regularity Theory.

Yet my disagreement with the Armstrong-Dretske-Tooley line on laws runs deeper than a disagreement over the significance of inductive inference. The burden of my argument thus far has been that a Humean account of laws is mistaken and that, when we think of laws in a classical context, the relations between properties looks to be far closer than simple accidental concurrence. But the classical context was crucial here — as we will see in the next section when we look to quantum mechanics. So what I would like to emphasize at the end of this section is that Humeanism is a mistaken view of natural laws. In the next section I argue that the Armstrong-Dretske-Tooley view is mistaken also, but not because it misrepresents laws, but because it misrepresents properties.[6]

II

The notion that laws of nature are a matter of relations between properties, whether accidental or otherwise, seems tailor-made for the metaphysical picture that is associated with classical physics. Properties like mass, velocity, charge, position, though complicated even in classical physics by relativity principles, have the character of *givens*, of values that are independent of the observer (modulo scaling). Of course early formulations of Special Relativity did confuse matters somewhat by speaking of velocities as being relative to *observers* but we now happily translate these anthropocentric references out of the physics by speaking instead of velocities as relative to *frames of reference*. This hiccup aside, physics up to the 1920s seemed to be progressing toward the ideal goal of a Lockean science: physics looked to be sorting out the genuine primary properties, factoring out the merely derived, the secondary, the composite, to give us a picture of the observer-independent physical world. To conceive of laws of nature as relations between properties is a natural concomitant to seeing physics as a well-designed sieve for excluding everything but the real properties of the world. Whether laws are thought of as Humean regularities or as some kind of necessities is really irrelevant from this point of view: they both share the classical *Weltanschauung*.

However this classical picture of laws as relations between properties ill-fits quantum mechanics, as we will see shortly. Even identifying the initial candidates for laws meets with problems once we step out of the classical context. This is because we are rather accustomed to meeting with formulations of classical physics in which certain simple relations are brought to the fore and made the basis for the whole of the science; Ohm's Law, Boyle's Law, Newton's laws of motion, these are all simple formulae that can be learned and applied once we know how to measure the relevant properties. To a very great extent learning classical physics is a matter of coming to understand these laws. The mathematics involved in the formulation of these laws has, with the passage of time, come to seem familiar and common-sensical: calculus, functions of real variables and second-order differential equations are as classical as the physics itself. Against this backdrop the laws stand out in relief as elegant theorems or axioms — exactly what they would have been in Newton's *Principia* and related works.

There is a clear pedagogical advantage to be had from singling out certain simple relations and calling them laws, particularly when they have their origin in an axiomatic presentation of a theory. When we come to Quantum Mechanics (QM) we find that we hear much less talk of 'laws' altogether. For someone to speak of the 'laws of quantum mechanics' would be rather strange: we would understand them to mean simply the entire mathematical and physical structure of quantum mechanics, and nothing less. Perhaps it is because there is no way that one could learn quantum mechanics by singling out a few simple relations and learning how to apply them — pedagogically, quantum mechanics confronts us with daunting prospects far beyond anything found in classical physics. Perhaps also there has not been the historical reliance on an initial axiomatic formulation — these have tended to come later and been the subject of dispute and revision.[7]

I mention these problems because, when we speak of the laws of quantum mechanics, it is not obvious what we ought to be speaking about: should we select out some simple relations and try to identify them with the laws, or should we be attending to the theory in its vast entirety. The answer, I think, is both and neither. We want to consider the theory in its mathematical entirety, but not in such detail that we need to be concerned with modelling actual situations. I need hardly add that in the space allowed this can be nothing more than a sketch — but it is still possible, even within these limits, to make some necessary points.

If a quantum system is undisturbed it will evolve deterministically according to the Schrödinger Equation; however, if a measurement is made on the system then it will undergo a sudden acausal collapse into an eigenstate of the operator corresponding to the observable that is being measured. What we have here —

absurd in its brevity — manages to encapsulate most, if not all, of the paradoxical features of quantum measurement. We will return to these 'paradoxes' later. What happens when we try to measure two observables at once? If the two corresponding operators do not share any eigenvectors then a collapse into one of the eigenstates of one of the operators means, *ipso facto*, that it cannot be in an eigenstate of the other. Thus, revealing an eigenvalue for one observable means that none can be revealed for the other.[8]

It is worth mentioning that the explanation for this phenomena is often misstated — as it is, in fact, in John Forge's article (this volume). When two operators **A** and **B** share no eigenvectors they are said to be complementary, to use Bohr's terminology. This is often equated, as it is by Forge, with the non-commutativity of the operators, as though it were the case that non-commutativity necessitated that no eigenvector of one could be an eigenvector of the other. This is wrong however. Non-commutativity only implies that *not every* eigenvector of **A** is an eigenvector of **B** and vice-versa — non-commuting operators are allowed to share some eigenvectors.[9] The upshot of this is that non-commutativity *does not* ensure that two observables can never be measured simultaneously — in spite of frequent claims to the contrary. When we come to the position and momentum observables the situation is complicated even further by the fact that their associated operators have no eigenvectors at all and hence no eigenvalues.

Failure of commutativity is a general condition of which the relation between position and momentum is a specific case, for these operators satisfy the canonical commutation relations (CCRs):

$$PQ - QP \leq - i\hbar \, I, \qquad \text{for } D(P) \cap D(Q)$$

These domain restrictions D(P) and D(Q) indicate that the operators related by CCRs are defined on only a subset, in fact a dense subset, of the Hilbert space vectors: outside of the intersection of those sets the CCRs are undefined.[10]

The fact that position and momentum satisfy the CCRs does not by itself ensure that position and momentum are not simultaneously measurable but another argument does guarantee that it is so. Consider an arbitrary wave function $\psi(x) \in L^2(\Re)$ in position space and a projection operator P_Ω which cuts $\psi(x)$ down to zero outside of some finite measurable set Ω. The new function has compact support and so, by the Paley-Weiner theorem, its Fourier transform is non-zero almost everywhere on the real line.[11] But the Fourier transform gives the momentum function and since this new function has non-compact support over the entire real line it cannot be an eigenvector of any momentum projection operator (except the trivial one that projects onto $\Re$ itself). In short any vector which is an eigenvector of a position projection operator could not also be an eigenvector of a momentum projection operator when the set Ω has finite measure. Or even shorter: any measurement of position, however coarse the localization might be, forces the momentum to be completely unlocalized — it has a non-zero probability of being any value at all. The simultaneous measurement of position and momentum must be a physical impossibility.

It is this argument that shows that a system may have a value for one observable but, of necessity, lack it for another, since it could not be in an eigenstate of both. Non-commutativity gives necessary but not sufficient conditions for this to hold. It is because of this that claims for the non-simultaneous measurability of two observables must be made in a piecemeal fashion, considering the nature of the particular operators in question. As it turns out position and momentum are a special case; for most pairs of operators we can only conclude that when a system is in a definite value state of one operator it is *possible* that it be in no definite value state of the other. Non-commutativity delivers this possibility and marks the divergence of quantum mechanics from classical physics. For some quantum systems, spin-one systems, the Kochen-Specker theorem provides a geometric proof of this non-classical nature. We have therefore, very strong reasons to believe that quantum systems need not possess definite values for every observable.[12]

The abstruse technicalities of quantum mechanics are liable to obscure the metaphysical message. To say that systems need not have definite values for particular observables is tantamount to saying that they do not have certain *properties*, in the classical sense. A system can simply lack a definite position, momentum, spin, energy, or what have you. But if systems can lack (classical) properties what remains of the characterization of laws of nature as relations between properties? The dispute between the Humean and the non-Humean as to whether nomic connections are accidental or necessary (in some sense) seems to have had the rug pulled out from under it: what is the point in discussing the nature of a relation when the relata have vanished?

This *problem of properties* in quantum mechanics could be described as 'fatal but not serious'. Or, to put it so as not to monger paradoxes, there is an aspect of this problem that is fatal to the hope of taking quantum mechanics straight, without revision, and yet the implications for understanding laws are not as dire as might first appear.

In brief, the problem for quantum mechanics itself, is that the collapse of a system into a definite value state (an eigenstate) is brought about by a measurement being made on the system. If we think of the system as being in an eigenstate as equivalent to its having a definite property corresponding to the observable being measured, then this definite property is conditional upon a measurement. The problem for quantum mechanics is that it is not obvious that we can cash out the notion of measurement in a non-anthropocentric way. It is not enough to think that the interaction between the measuring device — a large macroscopic instrument which is described, to within a good approximation, by classical physics — and the particles that are being measured, will by itself take one from superpositional states in the system to eigenstates. The joint system will be something that evolves according to the Schrödinger equation. Far from the measuring device collapsing the wave packet of the particle, the particle superposition will force the measurement device into a superpositional state, meaning that the pointer will be in a superposition of different pointer-readings.[13] Yet, not only do we not observe such pointer-superpositions, we have no idea of what it would be like to observe them. It was just this point that Schrödinger wished to make with his example of the superpositions of live cat/dead cat. The cat is here acting as a surrogate pointer-reading using its own mortality as the read-out device. The popularized moral of Schrödinger's cat, that consciousness is required to collapse the wave packet, is only a misleading way of putting the more serious point that quantum mechanics alone will produce macroscopic superpositions which we never observe; to bring this false account back into alignment with observation the observation itself is said to have miraculous properties: observation produces pointer-eigenstates where quantum mechanics cannot. Yet it is the height of *folly* to accept this as a straightforward consequence. The moral is not that consciousness can collapse the wave packet, but simply that the orthodox presentations of the measurement process cannot. We should not *believe in consciousness,* we should *disbelieve* in quantum mechanics — or at least the standard presentations of the measurement process that quantum mechanics employs.[14]

(Surely if consciousness really were able to collapse the wave packet it would not need a measurement device to be involved as well. What do the proponents think a geiger counter or a cloud chamber are? Altars? Magic Wands? Lightning-rods that can direct consciousness along its way? If consciousness were able to collapse the wave packet of particles directly then we would expect the Schrödinger equation to be nowhere applicable: we would, therefore, never have supposed it to be applicable! We would also expect individual consciousnesses to be involved in a bizarre cosmic tussle over which way particles should collapse. With this we are already a long way from the tranquilizing New Age pablum of a reified consciousness!)

It is difficult to see that the measurement problem can be solved in quantum mechanical terms, that is, without the need to make some adjustments in the structure of the theory. It is not just that quantum mechanics posits two distinct processes — Schrödinger evolution for unmeasured systems and probabilistic collapse for measured ones — it is that there is no satisfactory way of cashing-out the notion of measurement in physical terms. We therefore have no real way of demarcating the two processes.

Whatever the solution to the measurement problem should turn out to be, it is not likely to return us to the classical concept of an unconditional property: there are simply too many arguments that stand against such a notion (eg. the Kochen-

Specker and Gleason theorems), arguments that are independent of the measurement problem proper. But even though we will probably remain forever exiled from the Eden of unconditional properties it is unclear (and will remain unclear until we solve the measurement problem) what the conditional properties are conditional *on*. All of this has a bearing on the question of the character of laws of nature — particularly, of course, on the question as to whether there are any probabilistic laws — but we do not have to solve all of these problems to say something worthwhile about laws. We just have to reorient ourselves away from the idea that laws are relations between classical (i.e. unconditional) properties.

In many respects the notion of a property in quantum mechanics looks rather like one of Locke's secondary qualities: it is only after the light interacts with a normal observer that it is seen as 'blue'; just so, it is only after a measurement is made that the particle has spin up. But the similarity is entirely superficial because the light is not *changed to blue* by being observed whereas the particle *is* changed from a superposition to a spin-up state. A better analogy for quantum properties is provided by the, now rather unfamiliar, wheel chart. In a wheel chart a disc of cardboard is placed on a central spindle and allowed to rotate inside a cardboard envelope; there are pictures or words or numbers on the cardboard inner disc and these show through in the various places where holes have been cut in the outer envelope. When a picture (say) shows through a particular hole then we have a situation that is equivalent to a quantum eigenstate. In general, however, there will be holes through which no picture is showing (i.e. the system is in no eigenstate) and complementary holes (if a picture is showing in one then there will be no picture in the other).

III

Seeing properties as conditional does not mean giving up realism: it does not mean 'that the moon is not there when you aren't looking at it'. The situation is rather the reverse: the classical properties turn out to be only a small subset of the states that a system can be in. Thus quantum mechanics offers us an *expansion* of our ontology, *not* an anti-realist contraction, contrary to the ideology of early quantum theorists like Neils Bohr. Quantum mechanics offers us superpositions and mixtures, states that go quite beyond the classical mechanical properties. Quantum mechanics gives us an *enriched* realism where we can clearly see the necessity of breaking the (already attenuated) link between realism and observation that was bequeathed to us by the British Empiricists.

Given this opening out of our physical ontology we should regard laws of nature, not simply as relations between properties, but as relations between states of all kinds. Some laws may well be relations between properties, while others are relations between states in general: since eigenstates are simply states among other states there is no reason to think that laws must only hold between them. Yet some laws may hold between eigenstates: the EPR correlations are a good example of law-like connections between conditional properties, yet they hold in virtue of the nature of mixed states that enter into composite systems. The Schrödinger equation gives the law of evolution for undisturbed closed systems and relates arbitrary (pure) states, not just eigenstates of observables. As we move up from the quantum level the superpositional states become less noticeable in general — we might call this the *petering-out principle* as we move toward the classical limit — and therefore our laws seem more and more to relate solely to eigenstates and thus to, seemingly, classical properties. Yet since classical systems lack the rich variety of quantum systems we would expect them to lack also the rich variety of their laws.

Thus far we have measurement conditionals in which properties are relegated to the status of the consequents of measurement processes and a suggested expansion in the relata of laws, but we have not looked closely at the nature of the conditionals themselves. When we do we see that these conditionals are not at all straightforward.

To make a start on this suppose that **A** and **B** are operators measuring observables A and B and for simplicity let us only consider discrete spectrum operators (we can move to the spectral projections of continuous spectrum operators to avoid the domain problems that afflict the latter). In saying that A and B are conditional properties we mean that one of the eigenstates a_i and b_i, respectively, of these observables will be present when a measurement is made on a system. If A and B are complementary then no measurement that produces eigenstate a_i of **A** could leave the system in any b_i of **B**. If **A** and **B** do not commute then it is possible that if the system is in an eigenstate a_i of **A** that it will be in no eigenstate of **B**. If **A** and **B** do commute then if a system is in an eigenstate of **A** it will also be in some eigenstate of **B**. Since classical properties are a subset of the quantum states — the eigenstates and their associated eigenvalues — and since quantum systems are in general only in those special states as a result of measurements being performed, we can say that in quantum mechanics properties are conditional. More precisely, they are the consequences of conditionals whose antecedent makes reference to a measurement being made and where the whole conditional is governed by probability. We might represent this as follows:

$$P(M(A\ \psi) \rightarrow \alpha_i) = 1/n.$$

This can be read as: there is a probability of 1/n that a measurement of observable A made on a system in state ψ will yield eigenstate α_i. Such conditional properties look as though they are here to stay and they represent a sharp left turn away from the Lockean picture of classical physics. If properties become conditional then they are different kinds of entities than those conceived by Armstrong, Dretske and Tooley.

Probability is the probability of truth. The measurement conditionals that are being assigned a probability are, therefore, propositions that have to be able to have truth values assigned to them. Yet as soon as we look closely at these conditionals we strike a problem: the problem of counterfactual definiteness — or, as it is more commonly called, *the problem of conditional excluded middle.*

It is a perfectly acceptable quantum mechanical statement that, if a measurement is made on a system that the result will be one of the eigenvalues $\alpha_1 \ldots \alpha_n$. To assert this is to say nothing more than that a measured system will end up in some eigenstate revealing some eigenvalue — something that is very nearly analytic in orthodox quantum mechanics. The problem arises when we pass from this kind of conditional to the following:

Either, if a measurement is made then the result will be α_1,
or, if a measurement is made then the result will be α_2,
or, ...
.
.
.
or, if a measurement is made then the result will be α_n.

Disjunctions of conditionals of this form, though they look superficially to be simply a restatement of the eigenvalue conditional above, are controversial — indeed they may each have no truth value while the eigenvalue conditional is both true and uncontroversial. Put simply '$M \rightarrow (A \vee B)$' can be true even though '$M \rightarrow A$' and '$M \rightarrow B$' are neither true nor false.

The problem is that the disjunction of conditionals cannot have disjuncts with definite truth values because these definite truth values would allow one to derive the Bell Inequalities. Having these conditionals possess definite truth values is tantamount to assuming that systems possess hidden variables. Or so says David Lewis.

This issue arises because a number of authors, Eberhard, Stapp and Herbert and Karush, have shown that Bell's Inequality can be proven by employing counterfactuals with definite truth values rather than hidden variables.[15] As Lewis notes: 'It is considered nice that we can get Bell's Theorem using just the counterfactual outcomes, instead of trafficking in hidden variables as traditionally conceived.' Though he adds 'though for my part I cannot tell the difference.' So, according to Lewis, the disjunction of conditionals is really just a disjunction of hidden variable assignments.[16]

But this equivalence is not at all obvious. If I say that 'if a measurement were made the result would be α_i', am I really saying that α_i is already the possessed value before measurement? Certainly when I say that if a measurement were made the result would be one of the eigenvalues α_1 to α_n, I do not seem to be saying that either α_1, or α_2 ... or α_n are already possessed. Perhaps then the counterfactual is ambiguous. On one reading to say that if a measurement were made the result would be α_1 is to imply that nothing else could be the result because that is the possessed value. On another reading it is to say that it is what would *in fact* occur, even though there are many other things that *could* happen. These other outcomes are entirely possible, perhaps equally likely, but had the measurement been made this outcome α_1 would have eventuated. On the first reading we have hidden variables and in the second we have something more harmless but also more mysterious. We have definite truth-values for statements about what would happen if something were done though that thing is never done and the outcome would be a matter of chance. No one but God would ever be in a position to assert such conditionals but surely there is no commitment to hidden variables here.

The point about the derivation of the Bell Inequalities is that on either interpretation of such conditionals an assumption that they are determinately either true or false seems to be ruled out. Bas van Fraassen's response to such conditionals is notably different to Lewis'. He does not think that they are simply hidden variables in sheep's clothing but they are nonetheless ruled out by principles of Lewis' counterfactual logic.[17] In particular Lewis has rejected conditional excluded middle even in ordinary (classical) contexts. 'If Bizet and Verdi were compatriots then they would both either be Italian or French' can be true even though it is neither true nor false that 'If Bizet and Verdi were compatriots then they would both be Italian (alternatively French)'.

But van Fraassen 's response also seems to be inadequate: he leaves it quite unclear whether the assumption of counterfactual definiteness means that very many more conditionals will lack values than lack them in the classical context. It is quite unclear therefore whether these new derivations of the Bell Inequalities go beyond the traditional derivations that employ hidden variables. Yet this is just what they seem to do. According to traditional presentations of Quantum Mechanics there will be many classical counterfactuals that will lack a truth value and many classically true counterfactuals that will be false. An example of the latter is provided by the two-slit experiment: if a particle were to go through a single open slit then opening another slit would not affect its trajectory. This is a classically true counterfactual that is false in quantum theory.

The issue of which counterfactuals Quantum Mechanics sanctions, which it rules false and which it rules indeterminate is of great importance to an understanding of the metaphysical problem of quantum mechanical laws for the simple reason that — as was noted at the outset — laws sustain counterfactuals. Indeed the sustaining of counterfactuals is the mark of modality in a law of nature. Thus we can get a handle on the quantum mechanical laws by looking at the circumstances in which counterfactuals are supported. As in the classical case, this is given by the full mathematical structure of the theory.

In the traditional presentations of quantum mechanics, laws are not relations between properties in the classical sense but relations between states, both mixed and pure. Once we reorient ourselves toward such states we will stop expecting quantum mechanical laws to conform to a classical metaphysics. When we do this we will probably find ourselves as far from Humeanism as ever — but perhaps no longer in quite the same direction.

The University of Sydney, New South Wales

NOTES

1 A.J. Ayer, 'What Is a Law of Nature'. Reprinted in *Philosophical Problems of Causation* (ed.) Tom L. Beauchamp, Encino, Dickenson, 1974. Ayer uses the term 'conventional' where I have used the term 'subjective'.

2 D. Lewis, *Counterfactuals*, Cambridge, Mass. 1973.

3 For the Principal Principle see 'A Subjectivist's Guide to Objective Chance'. Reprinted in Lewis' *Philosophical Papers II*, Oxford, Oxford University Press, 1986. Strictly, given the evidence or lack thereof, you may have a range of degrees of belief about the bias of the coin.

4 D. Hume, *An Enquiry Concerning Human Understanding*, Oxford, OUP, 3rd ed. p.35.

5 D.M. Armstrong, *What is a Law of Nature*, Cambridge, Cambridge University Press, 1983. pp. 52-9.

6 I have other objections to the Armstrong — Dretske — Tooley view besides those addressed here. For one I think that the view fails the counterfactual test that was mentioned in the opening paragraph — I think only some kind of essentialist theory can meet such a requirement. For another, I think that unless something more is said about the necessitation relation than is usually said, the view will collapse into the Humean regularity account. These arguments will be developed elsewhere.

7 To save any possible confusion, it should be made clear that by 'Quantum Mechanics' is meant the standard theory in its usual mathematical presentation. Thus, for me, the many-worlds interpretation, for example — because of its significant mathematical deviations — counts as a different theory rather than an 'interpretation' of the standard one.

8 I will assume that the reader is familiar with the elementary 'picture' of quantum theory, as it is found in the usual text book accounts. Useful background material is contained in R. I. G. Hughes' *The Structure and Interpretation of Quantum Mechanics*, Cambridge, Mass., Harvard University Press, 1989. For a briefer introduction the reader might also consider my critical notice of R. I. G. Hughes book which emphasizes those elements of the theory that are important here (*The Australasian Journal of Philosophy* June, 1994).

9 The correct statement can be found in E. Beltrametti and G. Cassinelli, *The Logic of Quantum Mechanics* (Reading, Mass. Addison Wesley, 1981) pp. 26-8.

10 For more on these domain restrictions see my 'Unbounded Operators and the Incompleteness of Quantum Mechanics', *Philosophy of Science*, 57, 3, 1990, pp.523-34.

11 For the Paley-Wiener theorem see M. Reed and B. Simon, *Methods of Modern Mathematical Physics Vol. 1: Functional Analysis*, New York, Academic Press, 1980, p. 333.

12 For an account of the Kochen-Specker theorem see Hughes, *op. cit.*, pp. 164-8.

13 Some may object to the term 'particle' here — on the ground that particle-like behavior is itself a result of the measurements that we make. I see no reason to be that scrupulous: 'particle' is a perfectly good term to apply to quantum entities — a judgement that, at any rate, has already been made for us by most working quantum field theorists.

14 Of course by saying that we should not 'believe in consciousness' I mean only that we should not believe in consciousness as an irreducble, non-physical entity, capable of acting on

physical systems from the *outside*, as it were: we should certainly continue to believe in consciousness as a psychological phenomenon.

[15] P. H. Eberhard, 'Bell's Theorem without Hidden Variables', *Il Nuovo Cimento* 38 B, 1977, pp.75-9; N. Herbert and J. Karush, 'Generalization of Bell's Theorem', *Foundations of Physics* 8, 1978, pp.313-7.

[16] D. Lewis, *Philosophical Papers Vol. II*, Oxford, Oxford University Press, 1986, p.182.

[17] Bas C. Van Fraassen, *Quantum Mechanics: An Empiricist View*, Oxford, Oxford University Press, 1991, pp.122-5.

JOHN FORGE

LAWS AND STATES IN QUANTUM MECHANICS *

The topic to be discussed here, laws in quantum mechanics (QM), is important to me because of the particular account of explanation which I subscribe to. I have called this account the instance view (Forge 1986a). It can be summarised as follows: Something is explained when it is shown to be part of, i.e., an instance of, some suitable pattern in the world. The instance view therefore falls under what has come to be called the ontic conception of explanation.[1] The project of developing the theory of explanation along these lines then comprises two main tasks: identifying suitable patterns and showing how 'things' fit into them. The former is perhaps the more important, and the more testing, for the philosophy of science, for it is not very helpful to take "suitable" to mean "explanatory". We need some independent characterisation of what makes a pattern suitable for the purposes of explanation.

The problem of deciding which patterns are explanatory can be dealt with, or rather shifted, by taking them to be laws of nature. This shifts the problem because it is now necessary to determine which patterns in nature are laws. There are a number of interesting accounts of laws of nature available that do not presuppose that patterns are laws because they explain. An account such as Armstrong's (which holds that laws are relations of 'contingent necessity' between universals) can be stated without making any presumption about the role of laws — more of this later. So the instance view can be stated in a non-circular manner. But to apply and articulate it, to show that it is worth its salt, we need some examples of laws from science to work with. In classical physics these are easy to come by, but alas this is not the case in quantum mechanics.

Quantum mechanics presents the philosopher of science with a host of problems; consequently it is a very interesting theory. The main problems centre on *interpretation*. The formal structure of the theory is quite clear, although there are several ways to express the formalism. One interpretation, the 'no interpretation' interpretation as it were, is that the theory is a means for codifying measurement results. This is the refuge of many practising physicists. Nearly all philosophers want some form of realist interpretation — even the notorious anti-realist Bas van Fraassen is no instrumentalist. Realists must, however, be ready to give up some of their cherished classical principles, like causal connectedness, locality, value-definiteness, etc., when they come up against quantum mechanics. Now the question as to what the laws of quantum mechanics are like is also a matter for interpretation: If the theory is underpinned by laws of nature which are independent of us, then we need to know more or less what it is telling us

P.J. Riggs (ed.), Natural Kinds, Laws of Nature and Scientific Methodology, 163–185.

about the external world if we are to have some idea about these laws. Of course, the overall question of the interpretation of quantum mechanics is a very big one indeed: I shall only be concerned with it to the extent that it bears on the topic at hand.

I am going to make use here of van Fraassen's modal interpretation (see Forge 1993b and van Fraassen 1991). I am going to use this interpretation, to be discussed in Section 4, to implement the idea that laws are constituted by relations between *states*. This will be done in Section 5. To set the scene, I will introduce states in Section 1 in the context of classical physics. Then in Section 2, I will show that a (straight-forward) reductive account of states is not possible in quantum mechanics if we are realists. In Section 3, I consider some options for understanding laws in quantum mechanics in terms of states and in the final section, Section 5, I consider one of the options, 'option 4' which is the one inspired by the modal interpretation, in more detail. I would stress that, like much else in this field, the outcome is rather tentative and speculative. Peter Forrest's attitude that we should speculate about quantum mechanics is, I believe, a healthy one, given that there is no interpretation that commands general assent. (See Forrest 1988, p.x.) Moreover, I will make a brief attempt to try to express the candidate laws in terms of the Armstrong account of laws of nature, although I would stress that the proposals to be made in Sections 2 to 4 do not stand and fall with their implementation in that style of metaphysics.

1. LAW AND THEORY

When we restrict our attention to 'elementary' or 'empirical' laws, then we can easily get the impression that these are relations between quantities. Boyle's Law, for instance, looks very much like a relation between the quantities volume, temperature and pressure; Ohm's Law looks like a relation between resistance, current and potential; and so forth. This is because measurements on suitable systems, like samples of gas, electrical circuits and so on, yield values for quantities, which we attribute to the systems, that can be predicted directly by the laws. The attribution of quantities to systems and the formulation of laws come together in the theory of measurement.[2] This then underpins experimental practice and is vindicated by the fact that, within the limits of error and idealisation, values predicted are on the whole observed.

Theories, on the other hand, seem to *contain* laws. Perhaps this apparent conflict can be resolved simply by realising that theories are classes of statements, including statements of law, and that these together describe rather more complex and pervasive laws of nature than do empirical laws. Thus in classical mechanics, the three laws of motion and the force laws together (seem to) describe a rather deep pattern or law of nature constituted by masses accelerating under forces. On this way of looking at the matter, laws are still relations between quantities, but the relations can get rather involved, as does the corresponding account of measurement, when we go beyond empirical laws. There is, however, an alternative position. It is that when it comes to theories, the relations in question are between *states*. Whether this is equivalent to, or reduces to, the notion that laws are relations between quantities is something to be considered.

A theory of physics has associated with it certain *spaces* of states. These represent the domain of the theory. Classical mechanics has, for example, associated *phase* spaces: for each n-body system there is a 6n dimensional space that represents all possible states of the system. I will say more about what it is for a phase space to *represent* possible states in a little while. For the moment we may note that there are all sorts of interesting and useful things we can do with the space. For instance, the quantities exhibited by the system can be defined as functions of the phase space dimensions or variables: that is to say, they are functions of position and momentum. Alternatively, quantities are often said to be mappings from the space into $\Re^n$. Although this is the way mathematical physicists like to talk about quantities (e.g. Piron 1976, p.7), simply assigning n-tuples of real values to points or states has no physical significance unless these are values on some *scale*.

The phase space $\mathbf{P_i}$ for a given system s_i is a *class* and hence it is susceptible to the usual set-theoretical operations. We could, for example, form the power class $\wp(\mathbf{P_i})$ and it turns out that certain of the subclasses of $\wp(\mathbf{P_i})$ are of interest. If the system is located at a given point in phase space, if it is in the state represented by that point, and if its state changes, then it will end up at some other point in $\mathbf{P_i}$. Let an *evolution*, or *development* or *motion*, of the system take place when one of these 6n variables changes in value. Thus after a development has taken place, the system will have changed its location in $\mathbf{P_i}$. The motion is represented by the class of points $\mathbf{C_i}$ 'visited' by the system during the evolution and this is an element of $\wp(\mathbf{P_i})$. Given that $\mathbf{P_i}$ is the class of all possible states of system, then $\wp(\mathbf{P_i})$ will contain all possible motions $\{\mathbf{C_i}\}$. Generalising, let $\{\mathbf{C}\} \in \bigcup_i \wp(\mathbf{P_i})$ be the class of all possible motions for *all* mechanical systems; now we can specify the (representation of the) law of classical mechanics, $\mathbf{L(C)}$, in terms of constraints on the class $\{\mathbf{C}\}$.

What I mean by a 'constraint' here is simply a condition that restricts $\{\mathbf{C}\}$, which narrows it down in some way. Unless there were some restrictions on the motions of mechanical systems, it would be difficult to see how to attach any sense to the idea of a law of mechanics. The work of classical mechanics would then be accomplished by what we normally take to be just the necessary preliminaries of identifying the space of states and defining the quantities, together with the injunction "anything goes" — or, rather, "anything goes anywhere, anytime". The formal structure of the theory is too rich, in that it can represent much more than the motions that real systems can and do undergo, which is in fact as it should be. Formulating constraints on $\{\mathbf{C}\}$ is thus the set-theoretical equivalent of formulating law statements.[3]

One such constraint is that certain continuity conditions must be satisfied: systems cannot skip over temporal and spatial intervals and their momenta must change smoothly. Another is that the start of a motion must be before its end: assuming each $\mathbf{C_i}$ to be a sequence of states, then the order of the sequence must respect the temporal order. Moreover, if the motions of any two systems start off from the same initial state p_0, not only will the order of the ensuing sequences respect the temporal order, but, provided there is no differential 'interference', the sequences will be the same. Which is to say that **L(C)** is a *deterministic* law. The notion of determinism can be understood much more clearly in terms of a group of *operators* $\mathbf{U}\tau$ on $\mathbf{P_i}$. Not only will this help to clarify the 'action of time' on the phase space, it will be useful in Section 3 when we come to the Schrödinger dynamics of quantum mechanics. For these reasons I introduce the idea of an evolution operator here.

Suppose s_i is in the initial state p_0 at time $t = t_0$, then at any future time $t = t_n$, its state is $p_n = \mathbf{U_n}\, p_0$, where $\mathbf{U_n}$ is the appropriate evolution operator.[4] This operator is a mapping of $\mathbf{P_i}$ onto itself, which is to say it associates each point in the space with another (unique) point therein. While $\mathbf{P_i}$ must, so to speak, respect the action of time, the elements of the space are not temporal in the sense that, once gone, they are past for ever. Many different systems can take on p_0, provided of course that they exhibit the corresponding values for the 6n variables. And this need not be the initial state of a motion, which would mean that we would need to relabel it and use another subscript. This consideration does not affect the action of $\mathbf{U_n}$, as this represent the action of *temporal intervals*, not absolute time.

We assume that $\mathbf{U_n}$ is a member of the group $\mathbf{U}\tau$ of operators parametrised by the real numbers. What his amounts is that for *any* real numbers x and y and all states $p \in \mathbf{P}$

$$\mathbf{U_x}(\mathbf{U_y}p) = \mathbf{U_{x+y}}p = p_{x+y}$$

and

$$\mathbf{U_x}^{-1}\, p = \mathbf{U_{-x}}\, p.$$

What the first condition says is this: Suppose we consider the effect of operator $\mathbf{U_x}$ on a state mapped by the operator $\mathbf{U_y}$, then these two transformations $p \rightarrow \mathbf{U_x}p \rightarrow \mathbf{U_y}(\mathbf{U_x})p$ is the same as $p \rightarrow \mathbf{U_{x+y}}p$. The significance of this can be seen if we go back to our earlier example of the system in the initial state p_0 moving to the final state p_n. The first condition tell us that if operator $\mathbf{U_n}$ is in the group, which we assume it is, then for any real numbers r and q such that r +

$q = n$, there exist $\mathbf{U_r}$ and $\mathbf{U_q}$ such that $\mathbf{U_n}p_0 = \mathbf{U_r}(\mathbf{U_q})p_0 = p_n$. Now consider the sub-intervals (t_0 — t_r) and (t_r — t_n); $\mathbf{U_r}$ maps the motion in the first sub-interval in the sense that any system starting out in state p_0 will end up in p_r and $\mathbf{U_q}$ governs the second sub-interval. But since the first condition states that x and y are *any* real numbers, these sub-intervals can be divided up again and again, each time governed by an operator in the group which determines that a system in the state corresponding to the start of the sub-interval will end up in the state at the end — that is what it means to say that the group is parametrised by the real numbers. I think it is evident that any system starting out in p_0 and ending up in p_n whose motion is governed by $\mathbf{U}\tau$ will therefore experience precisely the same motion. The second condition, less important for our purposes, implies that the same is true is we change the sign of the parameter.

I said that two systems that started out in the same state would end up in the same state provided that there was no differential interference. In mechanics we think of such interference as *forces*, and we recognise that if a force is switched on at time $t = t_r$, s_i will not end up in the same state as it would have done without the action of the force.[5] Thus the group $\mathbf{U}\tau$ can be thought of as governing or picking out motions that are force-free and accord with Newton's first law. However, the fact that a force acts during some or all of a motion does not mean that its deterministic character is compromised and nor does it mean that such motion is not governed by (picked out by, represented by) evolution operators. The most general way to do this would be to consider groups of operators for all possible force fields or potentials. A system in p_0 which is then 'dropped into' the force field $\mathbf{F_j}$ will thus not develop in accordance with $\mathbf{U}\tau$, but in accordance with the group $\mathbf{U_j}\tau$.

The class of allowed motions $\mathbf{L(C)}$ is therefore specified by (very many) groups of evolution operators that fix possible phase space motions. Those constraints would be sufficient were we only interested in $\mathbf{s_i}$ — and provided that system did not resolve itself into component parts. But once other systems are taken into account, the effects of motions on one another must be considered. For instance, when two masses collide, their motions become correlated in accordance with the conservation laws. Again, the operator formalism could be pressed into service and we could introduce operators acting on the *products* of phase spaces for individual systems. For example, if $\mathbf{s_i}$ and $\mathbf{s_k}$ collide and then move apart, we may think of them thereafter as the complex system $\mathbf{s_{ik}}$ and use evolution operators on $\mathbf{P_i} \times \mathbf{P_k} \rightarrow \mathbf{P_{ik}}$. Henceforth, then, I will assume that the constraints that determine $\mathbf{L(C)}$ can be spelt out, for instance in terms of evolution operators. But how is this to be *interpreted*? What does it *represent*?

We could simply say that it represents all possible initial states of any mechanical system and all possible allowed motions from those states. This is already more than would be granted by the diehard instrumentalist. Representation, as understood by the instrumentalist, refers only to a basis for the prediction of observable properties, with no regard taken for any independently existing states, processes, etc., which connect up and mediate between measurements. Clearly we cannot be content with that viewpoint once we have evinced an interest in explanation. A realist, then, claims that states and motions of mechanical systems are real in that they exist independently of measurement and observation, and this claim focuses attention again on just what is being represented. For example, do we admit independently existing entities called forces represented by L(C) and the subclasses picked out by the various operator groups? I think we have a choice to make here. To admit forces as an independent category is to buy into a fairly rich variety of realism, favoured by those with a taste for causal explanation. I would remark that I do not share such tastes and hence have no brief to admit forces as a separate category. There is, however, no need to get embroiled in that issue here, for the interpretation of states is much more interesting.

If points in phase space are 6n-tuples of values, then surely states — real states, states 'out there' — are just collections of properties, properties that are in fact the degrees to which quantities like components of momentum and position are exhibited.[6] Moreover, it is convenient to collect these properties together as states because it turns out that they are sufficient for predicative (and explanatory) purposes: given the values of these quantities for s_i at t_0, we can predict what they will be at t_n. If a momentum component, say the x-component, is missing, then we will not be able to predict the value of *either* this component *or* the x-coordinate at any future time. On the other hand, we need no more information for predictive purposes than is supplied by the 6n values (cf. Margenau 1950, p.179). These values do not suffice to determine all mechanical properties; for example, the kinetic energy of s_i is not determined solely by its momentum. So we could add this value and use another state space, one with (6n + 1)-tuples. But this is not necessary. In the force free case, we can compute the kinetic energy of the system from its mass. That value does not change and hence once measured can be used at any time. When a force is acting, kinetic energy can be computed from a knowledge of the form of the force function. In this sense kinetic energy is not an independent variable. Now, if the realist is asked why these quantities, position and momentum, form a sufficient basis for prediction, then she may well reply that this is a consequence of the fact that they are actually related and that this relation constitutes a law of nature.

I will refer to the position taken in the previous paragraph as the *reductive* account of (real) states. It is reductive because it does not countenance any sense in which there are states over and above collections of properties and hence it denies that there is an independent category of states. The reductive account cannot be tested directly: it is not possible to set physical science to one side and see whether there actually are states, quantities and laws. Ontologies stand and fall on other grounds, in particular on whether they can give a plausible basis for our theoretical practice. An alternative position is that states of affairs are not just degrees to which quantities are exhibited; they have some separate and independent existence, *sui generis*. Moreover, if we were inclined to accept a non-reductive account of states, we may also be inclined to the view that laws are constituted by relations between states and not just relations between quantities. (In which case it would be necessary to distinguish between the law, namely the real pattern in the world, **l**(C) from its representation **L**(C) in the theory of classical mechanics.) On the face of it, this is an extravagant position, which will only appeal to those with a *penchant* for deoccamisation. Evidently states of affairs are very closely related to quantities and this is what is revealed in measurement: objects in states show certain corresponding values for position, momentum, etc. If this relation is not simply set-membership, as it is on the reductive account, then what is it? I will give some possible answers, and my preferred position, after I have sketched the non-reductive account of states in a little more detail.

It is as follows: there exist various sorts of states, mechanical states, thermodynamic states, electrodynamic states, etc., relations between these states, and individual physical objects which can take on, exhibit or instantiate these states. To provide some explication here, it would be prudent to appeal to some well-worked out metaphysical theory, perhaps a universals theory. But what attitude do we take to quantities? Are these attributes of objects because they are properties of states, second order universals, or are they properties of objects, or are they simply constructions?

My preference here is to regard values as possessed by systems *because* they have, or are in, certain particular states. It would, perhaps, be a little simpler to regard values as themselves properties of states, for then it will not be necessary to say what "because" means in the previous sentence. However, this alternative makes life difficult in QM, for it would then be necessary to attribute to states both occurrent properties *and* (something like) dispositions to exhibit those same properties. Now, of course, the only way we can come to know about states is through their properties. Thus we come to know what mechanical state a system is in by measuring, or inferring, its position and momentum. Although we need not endorse any naïve empiricism that eschews unobservables, there does not seem to be any good reason, thus far, why we should entertain *any* non-reductive account of states. When we turn to QM, however, we see that the non-reductive account is the only viable position and when we try to formulate some ideas about quantum mechanical laws we will see the advantage of the account I have just sketched.

2. STATES AND QUANTUM MECHANICS

QM is a well-established theory of physics and hence it associates a space of states with each physical system. The states in question are not, however, n-tuples of values of variables; they are vectors, or alternatively projection operators or density matrices, in Hilbert space — it will not be necessary to use the operator representation for states in this section but it will be necessary to do so in Section 4. A Hilbert space $\mathcal{H}$ is a vector space with representational capacity which allows us to express and codify all the measurement results that typify and characterise QM.[7] In particular, the Hilbert space representation gives rise to a statistical algorithm which allows the calculation of the probabilities of finding measurement outcomes. Since states are not represented as n-tuples of values, and I will spend some time in this section establishing this, it would seem that the main reason for adopting the reductive account of states is no longer compelling. If, as before, laws are thought of as classes or sequences of states, then, however we interpret the latter, they must be regarded as independently existing entities and QM cannot be thought of as simply an instrument for predicting the results of measurement. Can we do so?

Virtually all philosophers who call themselves realists would hope to answer this question in the affirmative, and it is certainly possible to do so. What is *not* possible is to interpret quantum mechanical states as something like the states of classical statistical mechanics. In general, the rules or algorithm of the theory do not allow us predict with certainty that a system will be found to have a certain property, a certain definite value for a given quantity. The statistical element of the theory cannot, however, be put down to our lack of knowledge. Rather, it seems that there is something fundamentally indefinite or indeterminate about most quantum mechanical states. If this is so, then a realist interpretation of states in QM cannot take these to be something like classes of 'degrees' to which all quantities in question are exhibited, *unless* QM is thought of as incomplete. This way out, which I will mention again, will not be taken here.

To make these matters more precise we need some details. If system **s** is in a state ψ which is an eigenstate of an operator **A** on $\mathcal{H}$ that represents the dynamical quantity A, then the algorithm can predict with certainty what the value of A will be found to be.[8] Since this always turns out right, it seems that we can say that s actually has the value in question. It seems that there is some close relation between being in state ψ and having value, say, a_1 of A, and thus far there is no reason to deny that (part of) being in ψ is having a_1. But if s is in ψ and if **B**, which represents quantity B, is incompatible with **A**, i.e., if the operators do not commute, then ψ is *not* an eigenstate of **B**. For an ensemble of systems identical to s, the algorithm gives the correct statistical distribution of B-values and in this sense QM gives the probability that s will exhibit a given B-value.

It can do so as a consequence of the fact that associated with each such operator there is a class of vectors, the eigenvectors of the operator, which *span* $\mathcal{H}$. What this means is that any vector in $\mathcal{H}$ can be written as a linear combination of vectors of, say, **B**. For instance if ψ is an eigenvector of **A**, which means it cannot be an eigenvector of **B** because these two operators are incompatible, then it can be expressed, for example, as $c_1\Omega_1 + c_2\Omega_2$, where the c's are (real-valued) co-efficients and the Ω's are eigenvectors of **B**. The square of the relevant co-efficient is the probability that the B-value corresponding to the B-state in question will be found on measurement. If **s** is in ψ, then it is said to be in a *superposition of states* with respect to **B**.

The received view is that in such circumstances B-measurement forces a discontinuous change or *collapse* in the quantum state $\psi \rightarrow \Omega$, where Ω is an eigenstate of **B**.[9] Thus, further measurements of B always give the same value, unless some A-measurement is interposed. If **s** is subject to alternate A- and B-measurements, then a range of different values for both quantities will be found. It seems that measurement does not merely reveal what values **s** has, it changes these values — measurement in QM is not passive. This prompts the question: When **s** is not in an eigenstate of quantity A, does it have *any* A-value at all?

Suppose, again, that there is an ensemble of **s** systems all with the same A-value a_1. Then a B-measurement is made and all systems giving the same B-value, say b_1, are subject to another A-measurement. It turns out that all manner of A-values are found, one of which may be a_1. We must acknowledge that measurement is faithful, at least in some sense: that is, it must reveal *something* about **s**. With regard to A-values, this could be either **s**'s A-value before measurement, after measurement, or both. The received view, noted above, is that A-measurement *leaves* **s** in the a_1-eigenstate, which is conveniently written as $|a_1\rangle$ in the Dirac notation. But what about beforehand? If all the selected **s** systems in identical $|b_1\rangle$ states all 'had' a_1 as well, then they would be also in eigenstates of **A** perhaps $|a_1,b_1\rangle$, which conflicts with the hypothesis that **A** and **B** are incompatible. So if we wanted to argue that B-measurement did not in fact change the value a_1, we would have to deny the theorem of QM which asserts that states can be represented by non-commuting operators (i.e., that there are no dispersion-free states, Hughes 1989, p.79). Suppose, then, that the selected **s** systems had a whole range of A-values before measurement. But measurement should now reveal *these* values, and not change them to a_1. No one is an operationalist any more, but it is difficult to make sense of the notion that something can be said to actually have a value for a quantity and at the same time it is allowed that measurement, presumably our best measurement, will change that value in an entirely unpredictable way.

The assumption that systems always have definite values for all quantities is sometimes called (not surprisingly) that of *value-definiteness*. It is generally agreed that this assumption cannot be maintained unless some well-established experimental results are denied, or unless QM itself is rejected as incomplete. Giving up value-definiteness is one of the more palatable responses to Bell's theorem (Cushing 1989). Giving up this assumption is also the best response to the theorem of Kochen and Specker (Redhead 1987, chapter 5). The considerations mentioned in the previous paragraph do not carry the rigour of these theorems, but they show in a straightforward way why the value-definiteness must be given up.

We must, however, be quite clear on just what is being rejected. It is that the Hilbert space vectors cannot be taken to represent real states that comprise definite 'degrees' for all dynamical quantities. And this is just what I understand the reductionist account of states to maintain. For on that account, states are just classes of properties which are selected as a convenient basis for predicting and explaining other properties. Quantum mechanical states fulfil this role, but they do not do so because they are simply classes of properties. We have seen that although ψ is a basis for predicting b_1, it does not *contain* individual values for all observables: the eigenstate $|a_1\rangle$ does not have b_1 as a member. Indeed $|a_1\rangle$ is not equivalent to $\{a_1\}$. This does *not*, however, mean that we cannot admit the existence of real states that are in some way, or are in some sense, indeterminate. Nor does it mean that we cannot draw a distinction such as that introduced by van Fraassen between *value states* and *dynamical states* (van Fraassen 1991, p.275). Let us adopt van Fraassen's terminology and refer to the real states of affairs associated with quantum mechanical systems and represented by Hilbert space vectors as "dynamical states". Is it possible to come up with some account of laws as sequences of (dynamical) states in QM?

3. LAWS AND QUANTUM MECHANICS

Hilbert space, like phase space, contains representations of all possible states of a quantum mechanical system. And quantum mechanics, like classical mechanics, places constraints on the way in which any given state can evolve. This last remark is not at odds with the probabilistic character of the theory. In the first place, it is not the case that QM always constrains the state to evolve probabilistically. In fact this only occurs, according to the received view at any rate, when measurement takes place. The principles governing such circumstances are called the *von Neumann dynamics*. When left to itself, i.e., when the system is not measured, its state develops deterministically (recall Section 1 and the discussion of evolution operators) and such processes are governed by the *Schrödinger dynamics*. It is certainly very strange that there should be two quite different sorts of fundamental processes in the world, and especially so since the former only came about when people began to perform measurements! Thus, even though physical processes would have occurred that have everything in common with a measurement except the presence of the observer, for example, photons colliding with massive objects and modifying the state of the latter, these would not have become measurements until a human observer came on the scene and devised a way to comprehend the information transferred to the massive object from the proton. Some philosophers, van Fraassen among them, have denied that there are two dynamics and have rejected the received view that collapse occurs on measurement.

I will now proceed as we did in Section 1 and come back to van Fraassen's interpretation in Section 4. So, let us once again form the power class $\mathcal{P}(\mathcal{H}_i)$ which gives the class of all 'evolutions' or 'motions' in the space $\mathcal{H}_i$ that represents system s_i.[10] Let $Q_i \in \mathcal{P}(\mathcal{H}_i)$ be the subclass picked out by the group(s) of evolution operators[11] and let all the sequences of dynamical states for *all* systems in the scope of the theory be represented by **L(Q)**. The suggestion, then, is that (a suitably refined version of) **L(Q)** be taken as the representation of a law of nature **l(Q)**.[12] This is (perhaps) not too controversial, given that we are willing to accept non-reductive accounts of states. Problems arise when we take the von Neumann dynamics into account. If a system **s** in state $\psi(t_1) = |a_1\rangle$ is subject to a B-measurement, then on the received view there is a discontinuous 'collapse' onto an eigenstate of **B**. Call this $\psi(t_2)$. $\psi(t_2)$ need be nowhere 'near' $\psi(t_1)$ in $\mathcal{H}_i$; in particular, if B is a continuous-valued quantity, like momentum, $\psi(t_2)$ could be very 'far' indeed from $\psi(t_1)$.[13] This is, of course, not at all deterministic: it is not governed by anything like **L(Q)**.

There are four main options that we can consider with respect to the question of laws and measurement interactions:

1. Deny that measurement interactions are lawful.
2. Deny that measurement interactions are lawful at the level of individual processes, but affirm that they show statistical regularity at the level of ensembles.
3. Maintain that measurement interactions are lawful at the level of individual processes.
4. Deny that the dynamic state ever collapses and hence affirm that **L(Q)** holds in all circumstances.

Option 4 leads to a metaphysical version of van Fraassen's modal interpretation. It is not without its difficulties but, as already mentioned, it is the one I shall opt for.

Option 1 is the easy way out, but it should be rejected straight away. There is surely no good reason to take the Schrödinger dynamics seriously and admit that this is underpinned by a law of nature whilst not doing the same for the von Neumann dynamics. QM has a very successful statistical algorithm which gets measurement results right every time, and a realist about the Schrödinger dynamics surely cannot claim that this is pure luck. We should be clear about the difference between options 1 and 4. The former accepts that there are two dynamics but denies that one is underpinned by a law. The latter denies that there are two dynamics. Options 2 and 3 are much more serious contenders. I will spend more time criticising option 2, as option 3 has some points in common with my preferred alternative, option 4.

For those who deny that there can be objective single-case probabilities, option 2 is the only alternative. However, if there are laws in QM that do not operate at the level of individual processes, then it seems that we have to admit *ensembles* as real entities. An ensemble is normally thought of as a 'conceptual entity' constituted by (a very large) number of systems in the same state. In classical statistical mechanics these are assumed to have different initial conditions — the initial conditions of the real system modelled by the ensemble are not known — and the theory gives the statistics of the evolution of the ensemble. In QM the assumption that identical systems evolve differently because of different initial conditions is equivalent to the assumption of hidden variables. We do not make such an assumption here, so we take the ensemble $E(a_1)$ to be a class of systems identical to **s** in $|a_1\rangle$.

Is it really necessary to introduce ensembles in this way? Suppose we did not, suppose we took option 2 as saying that there is a law of nature underpinning QM that is manifest directly in the actual distributions of values found in classes of measured systems. Then QM would be false. The distributions predicted by the statistical algorithm will only turn out exactly right in arbitrarily large classes, namely for ensembles. But there is another position that we need to consider and it is this: Ensembles are not real classes of systems but nevertheless they are what the statistical algorithm of QM is really about. The probabilities are then truly statistical, reflecting the proportions of systems in $|a_1\rangle$ that give, say, b_1 on measurement. The statistics of finite classes of real systems would then approximate those of the ideally large ensembles. This sounds attractive, but it is necessary to further interpret the claim that probabilities correspond to the statistics of ensembles, for if ensembles are not real entities then their components cannot in fact be measured. Perhaps this should be understood in the sense of a subjunctive conditional: if only there were a sufficiently large class of replicas of **s**, then the statistics of measurements thereon would reproduce exactly those of the algorithm. I confess to preferring a realist account of laws. Nevertheless, we will see now that there are more problems associated with option 2 which cannot be avoided by adopting this idealist interpretation of laws.

After a B-measurement, **s** will be found is some B-state. The ensemble $E(a_1)$ will be transformed into a collection of B-states with the distribution specified (more or less) by the statistical algorithm. The law must therefore involve some relation between ensembles and collections. It also seems that it must include some reference to measurement. One of the 'problems of the two dynamics', for those who do not subscribe to the Copenhagen interpretation, is to characterise measurement in such a way as to keep out the observer. If measurements were some microscopic-macroscopic interaction, such as when a photon collides with something big relative to itself, then such a characterisation would not have to refer to the observer. The state would always collapse in those circumstances and measurements would be distinguished as those cases in which we are able to come to know about the result of the collapse. However, on option 2 it would be necessary to distinguish those macroscopic interactions that produced B-states, from those that gave rise to C-states, etc. *If* all this could be done, then the (ideal) law would include various three place relations between ensembles, macroscopic objects and collections. Unfortunately, it has proved very difficult to draw the necessary microscopic-macroscopic distinction — this is not the place to go into those details (Albert 1992, pp.83-84).

One might try to get round this problem in the following way: System **s** will have a history — from cradle to grave, so to speak — and this could be told in more than one way. For instance, if **s** starts life in $|a_1\rangle$, we could chart its history in terms of the interactions it suffers; some of these might bring about the smooth Schrödinger-type evolution, while others might bring about collapse. Alternatively, the history could be told in terms of the successive states which **s** has. For certain times in its life, **s**'s state will develop smoothly, at others there will be abrupt change. For this history to be lawful, then it would need to be in accordance with **L(Q)** for the periods of smooth change. For the discontinuities, we might try to encompass all the possibilities in the following manner: Before each discontinuity, **s** belongs to an ensemble, such as $E(a_1)$ and afterwards it belongs to some collection, for example a collections of B-states (*and* to an ensemble E(b)). But to take care of the possibility that it might be in a C-state, or a D-state, or,..., it will be necessary to relate $E(a_1)$ to all the corresponding collections. This would, to say the least, be a little inflationary. I do not hold out much hope for option 2.

Let us move on to option 3. This corresponds to the *propensity* interpretation advocated by Popper, Maxwell and a few others. It is well explicated by Forrest, who suggests that propensities can be thought of as akin to *dispositions* of systems to respond in certain ways on measurement (Forrest 1988, pp.109-114). Thus the probabilities given by the statistical algorithm for a system in $|a_1\rangle$ to change to, say, $|b_1\rangle$ measures the 'strength' of its propensity to do so. Given that we want to constitute laws in terms of states, it seems that we must take propensities to be properties of states.

The law(s) that underpin the von Neumann dynamics will then be something like states of affairs, namely, the states of affairs of dynamical states having certain dispositions to change into other (incompatible) states. A system in $|a_1\rangle$ will be disposed to take on various B-states when subject to a B-measurement. If it is subject to a B-measurement, then one of these dispositions will become manifest in the sense that **s** takes on some B-state. To confirm what the statistical algorithm tells us about what the propensities are, we would need to perform a large number of experiments with replicas of **s**. But the law which underpins the statistical algorithm does not have to be constituted by ensembles, collections of states, etc. This, it seems to me, is the great advantage of option 3 over option 2. But option 3 is not without its difficulties.

If we agree to accept that propensities are (something like or something to do with) dispositions 'possessed' by states, we must ask what exactly are they dispositions for. A system **s** in $|a_1\rangle$ is disposed to take on various B-states on B-measurement. Moreover, **s** has these dispositions whether or not it is ever measured. There is nothing peculiar about some object having a disposition that is never manifest: glass is still fragile even if it is never broken. And we need not subscribe to the Logical Empiricist line that dispositions only make sense if they can be described by sets of reductions sentences. But notice that **s** has these dispositions until B-measurement, when it is then suddenly disposed to take on various A-states on A-measurement. When in $|a_1\rangle$, **s** is not disposed to any degree above zero to take on any other A-state. Yet B-measurement wipes out at a stroke all dispositions to take on a B-value on further B-measurement save one, namely that which corresponds to the B-state that it is in, and replaces them with a whole range of dispositions to take on various A-values. It is not immediately clear what sort of law might capture these sharp changes in propensity. If it is replied that propensities are associated with states and that sharp changes in the former reflect sharp changes in the latter, then the question is shifted to one about changes in state. This is where we came in! There seems no escape for the problems associated with the von Neumann dynamics. Perhaps a no-collapse interpretation of QM, such as the modal interpretation, might be an attractive alternative?

4. THE MODAL INTERPRETATION

The modal interpretation, on which we base option 4, denies that there is any collapse on measurement. Thus at all times the dynamic state develops in accordance with L(**Q**). We need some exposition of the modal interpretation before option 4 can be fully appreciated. To this end I adapt a schema from van Fraassen which makes clear the difference between his position and the received view:

$$\mathbf{U}\text{: IN, } \psi_0 \rightarrow \psi_1\text{, OUT}$$

$$\mathbf{U}\text{: IN', } \psi_0' \rightarrow \psi_1'\text{, OUT'};$$

here ψ_0, ψ_1, etc., are the initial and final states of system **s** which has undergone a particular evolution described by the operator **U**. Let IN and OUT be the initial and final states of the surroundings in which the evolution takes place; for instance, these could comprise a measuring system **m** connected to **s** at the beginning of the evolution. Suppose this is indeed the case, then IN will be the 'ready state' of **m**, **U** will describe the interaction of **s** and **m**, and OUT will be **m** specifying some value for a quantity on **s** by means of its pointer observable.

There are now two different ways in which determinism and probability could enter the picture:

(I) Final state ψ completely determines OUT. So if OUT $\neq$ OUT', then $\psi_1 \neq \psi_1'$. Probability pertains to the process: $\psi_0 \rightarrow \psi_1$ conditional on IN.

(II) Initial 'state' (IN,ψ_0) completely determines ψ_1. So if $\psi_1 \neq \psi_1'$, then (IN,ψ_0) $\neq$ (IN',ψ_0'). Probability pertains to OUT, given ψ.

(I) is the received view: ψ_0 collapses onto ψ_1 when the measurement takes place. But on (II) there is no collapse: (In, ψ_0) completely determines ψ_1.

Van Fraassen then refines (II) by introducing the distinction that we have already noted above, between the dynamical state and the value state. He identifies ψ as the dynamical state of **s**. OUT is assumed to be a faithful record of the value state at the end of the process, but it is not fully determined by the ψ_1 — in effect, OUT is the value state at the end of the interaction.[14] IN is not, however, the initial value state. The state (IN,ψ_0) is in fact the dynamical state of the composite of **s** and **m**.[15] The final state of **s**, i.e., ψ_1, which can be 'disentangled' from the composite, for example when **m** is disconnected, is determined by the evolution of this initial composite state in accordance with the operator **U**.

The anti-realist van Fraassen holds no brief for metaphysics, so it is up to those who would use his modal interpretation for their own ends to supply their own. We have accepted already that dynamical states are real states of systems and that they always evolve in accordance with **L(Q)**. There is only one sort of dynamics in QM, according to the modal interpretation, and in this respect it turns out like classical mechanics. There are, however, two sorts of states, dynamical states and value states, and we must see just how they are related, for it may turn out that there is (are) some lawful relation(s) between the two. To begin, we may look at what van Fraassen has to say about the differences between them.

Van Fraassen uses the following terminology: [A,a_1] stands for the *statement* that the dynamical state, namely $|a_1\rangle$, is such that measurement of A on **s** must give a_1, and "must" means probability = 1. If **s** is in $|a_1\rangle$ then [A,a_1] will be true, and it is $|a_1\rangle$ which 'makes it true' (van Fraassen 1991, p. 275). $\langle$A,$a_1\rangle$ stands for the statement that **s** actually has value a_1, regardless of whether it is measured or not. It is the value state, which I will write as (a_1,...), that makes this statement true. Moreover if a_1 is the outcome of an A-measurement on **s**, we assume that the value state of **s** is (a_1,...) — such a value state may comprise other values and hence we should not render it as (a_1). Van Fraassen therefore acknowledges that dynamical states and value states are both real. But they are not the same. Let ψ be the dynamical state of **s** ($\psi \neq |b_2\rangle$) then it is possible that even though [B,b_2] is false, **s** has b_2. It is this possibility that dramatically illustrates that the modal interpretation is a non-collapse interpretation. But now we need to address what seems to be a difficulty with the modal interpretation and in so doing we will become clearer on just what is novel about it.[16]

If $(.,c_3,..)$ is the value state of s and if $\psi \neq |c_3\rangle$, then is not QM incomplete in the sense that s has a property which the theory cannot predict? Namely that s 'really has' c_3, yet the statistical algorithm only gives a probability less than 1 for (c_3). And why stop at c_3, why not restore value-definiteness and have complete value states $(a_1,b_2,c_3,d_4...)$? There is, however, a difference between QM being incomplete in the sense that it cannot *predict* the results of measurement and in the sense of not having the capacity to *represent* something that can actually be found out by measurement. QM is an irreducibly probabilistic theory and so it cannot predict all individual measurement outcomes. In Section 2, when our thinking was conditioned by the received view, we assumed that if s was not in an eigenstate then it had no value for the quantity is question. Value-definiteness was rejected because there are no dispersion-free (dynamical) states. Van Fraassen is telling us that if we measure C and find c_3 we cannot necessarily infer that s is in $|c_3\rangle$. But if QM cannot represent this state of affairs, then it seems that it is incomplete in a way that might well make us unhappy with the modal interpretation. For this is not incompleteness about predicting the results of measurement but about inference from measurement to dynamical state. On the received view, with its two dynamics, measurement brought with it collapse, in accordance with process (I) above. Thus, we always knew what dynamical state s ended up in.

To come to grips with van Fraassen's account of this matter, we need to understand what happens to the state after measurement. To this end, it is necessary to distinguish *pure* states from *mixed* states. This is most easily done in terms of projection operators. Let $\mathbf{P_T}$ be a operator and $\mathbf{T}$ a subspace of $\mathcal{H}$, then $\mathbf{P_T}$ is a *projection* operator, if, for all $y \in \mathcal{H}$, $\mathbf{P_T}y = py$, $y \in \mathbf{T}$ and p is number between 0 and 1 inclusive. One-dimensional operators $\mathbf{P_x}$ are particularly important because they can be used to represent pure states. As the name suggests, these operators 'project' onto a subspace containing (all scalar multiples of) a single vector: thus $\mathbf{P_x}y = x \bullet y.x$, where $x \bullet y$ is the inner product of the (normed) vectors, which again has a value between 0 and 1. If vectors can be used to represent states, then, since there is a one-dimensional projection operator (projector) for every vector, the latter can also be used for this purpose. It is often very convenient to do so. Now suppose we have a weighted sum $D = \sum_i w_i \mathbf{P_i}$ of projectors and that it is not the case that some $w_j = 1$ (i.e., that D does not reduce to a single projector), then D is said to represent a *mixed* state or *mixture*. It is not very easy to explain in an intuitive fashion what the difference between pure and mixed states is. I think it is better here to just say how they are differently represented and leave it at that. It used to be thought, and it is still part of the received view, that there are really only pure states and that mixed states were just a reflection of our ignorance of which pure states systems end up in after measurement. This *ignorance interpretation of mixtures* has been discredited (Beltrametti and Cassinelli 1980, pp.9-11).

Mixed states are important because, in general, **s** will end up in a mixed state after it is decoupled from the measuring instrument **m** (Beltrametti and Cassinelli 1980, pp.66-67). It is those circumstances in which van Fraassen allows that s might be found to have a value that its dynamical state does not 'force' it to have. One might wonder how this could *ever* be the case: if ψ is a mixture of projection operators which represent eigenstates, then its (real) dynamical state is some mixture of pure states and so we might wonder how there could be *any* statement like $[A,a_1]$ could ever be 'made true' by ψ. To answer this question fully we would have to go into the details of the statistical algorithm for mixed states, and this would take us too far afield from our present concerns. Suffice it to say here that for the modal interpretation to work, it must happen that ψ makes some such statement true.[17]

Value states do in fact have a representation in the formalism of QM, according to the modal interpretation. It is this: if ψ is the dynamical state of **s** and is a mixed state, then the value state of **s** is represented by some pure state x *possible relative to* ψ. x is possible relative to ψ if whenever ψ makes statement [..] true, so does x (van Fraassen 1991, p.140 and p.281). In other words, whenever the statistical algorithm gives a probability equal to 1 that a certain outcome will be found in ψ, it also gives this result for x. This definition allows for x to give probability 1 to certain other outcomes which are *not* given this probability by ψ.

For instance, x is possible relative to ψ if both make true $[A,a_1]$ and $[C,c_3]$, and this corresponds to the value state $(a_1,c_3,..)$. But x might also make true $[D,d_4]$. If so, then the value state of **s** is in fact (a_1,c_3,d_6). Van Fraassen claims, though I do not think he proves, that all the components of ψ, i.e., all the $\mathbf{P_i}$, are possible relative to ψ (van Fraassen 1991, pp.161-163).

Assuming that the only candidates for the representatives of the value state of s are the components of ψ, then, as van Fraassen remarks, it is as if the ignorance interpretation of mixtures is true. And this is indeed interesting.[18] It follows, on this reading, that if ψ is pure, then only ψ is possible relative to ψ. For example, if s is in $|a_1,c_3\rangle$, then only (a_1,c_3) is possible and all other value states have probability zero. Van Fraassen also remarks that if measurement reveals a value state represented by a pure state possible relative to the (mixed) dynamical state, then this constitutes the transition from the possible to the actual, and this is where the modal interpretation gets its name. It is, I think, clear that this interpretation does not imply that QM is incomplete as far as pure states are concerned. But once s is measured, it can end up in a mixed state, and this can open up a range of possible value states. But it is only if the ignorance interpretation of mixtures really were true, as opposed to seeming to be true as far as the representation of value states is concerned, that the modal interpretation would be open to the above charge. For in that case, measurement of the value state should enable us to infer the unique pure dynamical state that **s** is really in. Since it is not in any such state, the fact that we cannot decide (probability = 1) which one it is in is not problem. I conclude that the modal interpretation is worth taking seriously.

5. STATES AND VALUES AND LAWS

On the modal interpretation there are, as we have just seen, real value states and real dynamical states. It is necessary now to consider the relation between them, and in particular to consider whether it is a lawful relation. Recall that if the dynamical state ψ of **s** is pure, then it has the value state made true by this state. In other words, as above, **s** has the values which correspond to the eigenstate $\psi = | a_1, c_3, \ldots \rangle$, namely those of the value state $(a_1, c_3, \ldots)$. It does not follow that there is a lawful relation between pure dynamical states and value states, because the inevitable correspondence could be mere constant conjunction — assuming laws to be more than this. Nor does it follow from the presumption that **l(Q)** — we need to focus here explicitly on the law of nature as opposed to its representation — is a law about dynamical states that there is also a law about the relation between the *kinds* of state. But if we do adopt an overall realist standpoint with respect to QM, and if we are trying to develop an account of laws in QM, then I think it would be a little strange if we suddenly became bashful at this point and refused to postulate more laws. So, suppose there is a law **l(PV)** — the law of pure states and value states — constituted by the lawful correspondence between dynamical states and value states in accordance with the modal interpretation. But **l(PV)** only holds if **s** is not measured. If measurement leaves **s** in some other pure state $\psi \rightarrow \psi'$, which we saw is the exception, then we must assume that this happens smoothly in accordance with **l(Q)**, and that the result will be another instance of **l(PV)**. And throughout this 'motion', **s** will have associated with it corresponding pairs of dynamical and value states. Things are more interesting when the result is a mixed state.

It has been assumed here that the characterisation of relative possibility in terms of probabilities and in terms of components of mixed states pick out the same pairs of states. Mixed states are real dynamical states which systems can possess independently of our measurement practice. For instance, **s** could become coupled and then decoupled from (some object like) **m** accidentally, and as a consequence be left in the mixed state ψ. There will then be a whole range of pure states possible relative ψ and we must assume that **s** is in a value state corresponding to one of them. We cannot say that the 'transition from the possible to the actual' occurs next time **s** is measured, for that interaction will most likely put **s** into a new mixed state. Presumably the transition occurs at the same instant as the interaction with **m**. Now the relationship here between ψ and the value state (...) of **s** is not such that **s** *has* to have (...) and hence we do not have an instance of **l(PV)**. Let us therefore postulate **l(MV)** which is a *probabilistic* law governing the *propensity* of a system in mixed state ψ to have the value state (...). The proposal here is to ground relative possibility between mixed states and value states in the propensity of systems in those mixed states to exhibit certain value states. The statistics of the results of measurements can then be taken as evidence for (or against) **L(MV)**. What can be achieved if we adopt this proposal about laws in QM and how should it be developed?

It would appear that (real) dynamical states *explain* value states and hence explain measurement results. For instance, in answer to the question "Why does s exhibit a_1", it can be said that s is in $\psi_1 = |a_1,...\rangle$ and, moreover, it was previously in $\psi_0 = |....\rangle$ which evolved to ψ_1 according to **U**. I would add that these states of affairs are instances of **l(PV)** and **l(Q)** respectively. Now there is a considerable body of opinion which holds that if laws are to explain, then they cannot be mere regularities. In the context of the ontic conception this amounts to the requirement, noted above in the introduction, that not every pattern is explanatory. It will therefore not do to simply to say that laws are patterns in the world. Neither will to do to say that they are explanatory patterns. Some independent view of laws of nature is needed; and to this end I will make use of Armstrong's account to illustrate how the metaphysics might be worked out.

The idea, then, with regard to **l(Q)** is that dynamical states are *universals* and that relations of *contingent necessity* hold between them which constitute the law of nature. The relations are precisely those described by the Schrödinger dynamics, which is to say that they are ordered sequences of states. But unless we are willing to countenance uninstatiated universals, then **l(Q)** will be too rich. Presumably not *every* possible 'motion' is actually instantiated by some physical object at some time. Let **l(IQ)**, where **I** stands for "instantiated", be the collection of instantiated states. Then the representation of this collection, **L(IQ)**, belongs to **L(Q)**.[19] One could then regard scientific research as the attempt to pick out **l(IQ)** by exploring **L(Q)**. I think this is preferable to allowing uninstantiated universals.

All we have done so far is ascend to the level of universals and postulate a relation of contingent necessity between all instantiated sequences of states, namely between all the states that there actually are. This looks both arbitrary and empirically empty; it looks like a trick conjured up specifically to distinguish laws from constant conjunctions. I confess to have some sympathy with this objection, although I think it misses the point. The point is not to construct an account of laws that is in some way empirically testable, it is to produce one that has some explanatory power. One that can, for example, be used to explain the difference between laws and constant conjunctions. If the general theory of universals in which the account is embedded is successful on other counts, then its explanatory power is thereby enhanced: it gains by association. This is obviously not the place to rehearse the successes of Armstrong's general theory (see Armstrong 1978); I will simply assume that it has some. But can more be said about **l(IQ)**?

It is certainly possible to distinguish between individual dynamical states, like $|a_1\rangle$, from the 'generic' dynamical state. Armstrong would do this in terms of properties. He would distinguish the first order property "being in state $|a_1\rangle$" from the second order property "being a dynamical state". The latter is a higher order universal, which is to say that it is instantiated by (in this case first-order) universals, not particulars. This move is useful for the following reason: The extension of **l(IQ)** is the collection of classes $\{q_i\}$, where the q_i are sequences of real dynamical states. Suppose we ask how it is individuated. As a first step it can now be said that all and only instantiations of "being a dynamical state" are members of the union $\bigcup_i q_i$. Moreover, there may be lots of instances of "being in state $|a_1\rangle$", which will be the case if indeed lots of physical objects at some time take on this property. In which case some of the sequences q_i will be the same and some will be subclasses of others. The latter will eventuate, say, if two objects start out in state ψ_0 and are subject to the same evolution $\mathbf{U_t}$ until the first ends its life in ψ_t while the second lives on until time (t + 1).

All sequences q_i that are the same will instantiate the same law of nature which is constituted by a relation of contingent necessity between the corresponding universals. Thus **l(IQ)** turns out be the *collection* $\{q_i\}$ of (simple) laws of nature. It would make things neat and tidy if it were possible to bind this collection together. Armstrong himself attempted something similar in connection with quantitative (what he called functional) laws (Armstrong 1983, Chapter 7). In that case, the simple laws were relations between first order universals like "having resistance to degree i", "having a current of degree j", etc. Armstrong suggested that the members of a collection of relations between these first order properties could be instantiations of a higher order relation between the second order properties "having resistance", "having current". However, we have given up the idea that laws are relations between quantities. We are left with states and, apparently, with no way of dividing up states so as to produce a single higher order law. Anyway, there are difficulties with Armstrong's proposal (see Forge 1986b). It would therefore appear to a consequence of the non-reductive account of states that the view of laws to which it leads is a little untidy.

Turning now to **l(PV)** and **l(MV)**, we suppose that value states are also properties construed as universals. They *could* be construed as properties of dynamical states, and hence second order properties of physical systems, which would immediately resolve the issue of the substantive connection between the two types of states. This is a discussible proposition if ψ is pure, but if it is mixed it is not so attractive. Recall that if ψ is mixed then nevertheless **s** has a definite value state, a state that is possible relative to ψ. There will, in general, be many such possible (and incompatible) value states and hence these cannot all be (occurrent, possessed) properties of ψ. Here again we come up against one of the mysteries of QM: the lack of definiteness of the 'properties'. It does, however, seem necessary to introduce *propensities* here. The question is then to *what* do we attribute propensity?

For instance, suppose **s** is in the (mixed) state $\psi = \sum_i w_i \mathbf{P_i}$, then the *statistical* probability that it has the value state ($\mathbf{P_i}$) represented by $\mathbf{P_i}$ can be computed from $\mathbf{P_i}$ and ψ (it is Prob($\mathbf{P_i}$) = Tr $P_i\psi$). Should we attribute the corresponding propensity to the dynamical state ψ ? Then value states would be a kind of second-order property, but they would be dispositional properties that dynamical states have a certain given propensity to exhibit. This is a novel suggestion; but it is surely odd to postulate determinate properties as second-order. Surely, value states are properties of *physical systems*. In which case, they would be dispositions which system **s** has a certain propensity to exhibit *because* it is in ψ. We come back again to the problem of saying what is meant here by "because". I confess I can think of nothing better that to suggest that dynamical states be thought of as the categorical basis of the disposition. It would therefore bear a similar relation to the value state that the amorphous structure of glass bears to the disposition to break. But the relation is not exactly the same. For while glass always has the disposition, **s** has a propensity to exhibit a particular value state.

CONCLUSION

There is clearly a good deal more work to be done before the proposals made here can be accepted as an account of laws in QM: parts of Section 5 are, to say the least, speculative. However, I hope to have shown that the 'states approach' is worthwhile. Its strengths derive to a considerable degree from the modal interpretation which enables us to get round the problems posed by the two dynamics of the received view, and from the fact that it is compatible with the rejection of value definiteness and, as a matter of fact, with the rejection of causal connectedness. Moreover, if the states approach is good for QM, I do not see why it should not be extended to classical physics. Proponents of that approach can now plead more than a penchant for deoccamisation, they can argue for a unified account of laws in physics.

Griffith University, Nathan, Queensland

NOTES

* I would like to thank the editor for his kind invitation to write a paper for this volume and to the anonymous referee for helpful and perceptive comments on an earlier draft.

1 The instance view is, however, by no means the 'majority position' when it comes to the ontic conception. See my remarks in Forge 1993a.

[2] An excellent account of this is given by Brian Ellis in his *Basic Concepts of Measurement*, especially Chapter 8. Ellis now sees *Basic Concepts* as too conventionalist, see Ellis 1992, but, even if true, that does not detract from the account given of empirical laws in the book.

[3] Those who are familiar with the Structuralist approach will be aware that I am not using "constraint" in accordance with that tradition. To translate the present discussion into the language of Structuralism, it would be necessary, first, to identify **{C}** with the class of all partial potential models of classical mechanics and **L(C)** with class of models for the theory. "Constraint" is the term used in Structuralism for a special kind of condition that picks out a subclass of the partial potential models of a theory that corresponds to the class of models. See Balzer *et al.* 1987, pp.46-47.

[4] "Appropriate" in the sense that it gives the right answer. Much of the work of analytical mechanics is, essentially, the formulation of evolution operators.

[5] More precisely, it will not have the same motion, for it is possible that another force could compensate for the action of the first and hence that p_n would be the final state after all.

[6] This is not the place to worry about the sense in which position can be considered as a property, nor is it the place to worry about the relation between properties and quantities.

[7] By far the best book on the representational capacity of Hilbert space is R.I.G. Hughes, *The Structure and Interpretation of Quantum Mechanics*. Cambridge: Harvard University Press, 1989.

[8] A note on terminology: dynamical quantities or 'observables' are properties of systems whose values change as the system's state changes. Thus mass is not an observable in (elementary) quantum mechanics.

[9] What I mean here by "received view" is simply the position taken by nearly all philosophers and physicists, with one or two notable exceptions. It is the view that there are two dynamics in QM.

[10] We should actually taken the power class of 1-dimensional subspaces of $\mathcal{H}$, where vectors of the form $c\psi$, c is a complex number and ψ is normed to 1, belong to the same subspace. All the vectors in the subspace represent the same dynamical state.

[11] The operators U_t have the form e^{-iHt}, where H is the total energy operator, which is to say that it represents the forces acting on the system. For an more technical, and much more informed, discussion of evolution operators, see E. Beltrametti and G.Casinelli 1981, Chapter 6.12.

[12] As was the case in Section 1, we should distinguish the law as a real existing pattern in the world from its representation by means of the state space formalism of the theory. It will not, however, be necessary to make this distinction explicit again until Section 5.

[13] If the system is in a 'position eigenstate' then, according to the view expressed in the previous section, it will have no definite momentum. Nevertheless, QM allows us to calculate an expectation value for momentum measurement: this is a weighted average of values corresponding to an expansion of $\psi(t_1)$ in terms of eigenstates of momentum. To say that $\psi(t_2)$ is 'far' from $\psi(t_1)$ means that its coefficient in the expansion is very small.

[14] Van Fraassen spends a good deal of time going over the measurement problem and he gives us a detailed account of what goes on in measurement. In particular, he explains how system and measuring instrument couple and how information from the former is passed on the latter. See van Fraassen 1991, Chapters 7 and 8.

[15] As will become clear, the initial value state *is* represented by ψ_0, if this state is pure. If on the other hand ψ_0 is mixed, then this will be due to a previous encounter with a measuring instrument and the process described will have to be placed in the context of a series of measurements.

[16] The discussion becomes a little more technical for the rest of this section. The reader who is not interested in such niceties could skip to Section 5.

[17] Here's how it could come about: We cannot appeal to our simple examples of individual values like a_1, although I shall continue to do so in the text, we have to introduce another kind

of 'quantity'. Consider the operator Q^E whose eigenvectors $|F\rangle$ are such that $Q^E\,|F\rangle = F|F\rangle$, F $\subset$ E. We understand this mean that any system in state $|F\rangle$ will have a value for Q in the range of values E. Let $P^Q{}_{|F>}(E)$ stand for the probability that a system in state $|F\rangle$ has a Q-value in the range E. Evidently this probability is 1. Now suppose $\psi = \Sigma_i\, \mathbf{w_i P_{|Fi\rangle}}$, which is to say that it is a mixture of eigenstates of Q^E. $P^Q{}_\psi(E) = \mathrm{Tr}\,(\psi \mathbf{P_E})$, where $\mathbf{P_E}$ projects onto the subspace of eigenvectors of Q^E. It happens that this comes out as probability = 1. It is also the case that for any of the projectors, say $\mathbf{P_{|Fi>}}$, in ψ, this probability will also be 1. But providing there is some $|Fk\rangle$ represented in the mixture such that Fj $\subset$ Fk, then the pure state represented by $|Fj\rangle$ will 'make true' values not made true by ψ.

18 Moreover, it does *not* restore value definiteness because any state represented by Hilbert space vectors are not dispersion free.

19 A little more precisely, there will be a homomorphic mapping from **l(IQ)** to **L(Q)**.

REFERENCES

Albert, D.Z. (1992), *Quantum Mechanics and Experience*. Cambridge, Mass.: Harvard University Press.

Armstrong, D.M. (1978), *Universals and Scientific Realism*. Cambridge: Cambridge University Press.

Armstrong, D.M. (1983), *What is a Law of Nature?* Cambridge: Cambridge University Press.

Balzer, W., C.U. Moulines and J.D. Sneed (1987), *An Architectonic for Science*. Dordrecht: Reidel.

Beltrametti, E.G. and G. Cassinelli (1980), *The Logic of Quantum Mechanics*. Reading Mass.: Addison-Wesley.

Cushing, J.T. (1989), 'Background Essay' in J.T. Cushing and E. McMullin (eds.) *Philosophical Consequences of Quantum Theory*. Notre Dame: University of Notre Dame Press, pp.1-24.

Ellis, B. (1966), *Basic Concepts of Measurement*. Cambridge: Cambridge University Press.

Ellis, B. (1992), 'Conventionalism in Measurement Theory' in C. Wade Savage and P. Ehrlich (eds) *Philosophical and Foundational Issues in Measurement Theory*. Hillsdale, N.J.: Lawrence Erlbaum pp. 167-180.

Forge, J.C. (1986a), 'The Instance View of Explanation', *Australasian Journal of Philosophy*, **64**, pp.127-142.

Forge, J.C. (1986b), 'David Armstrong on Functional Laws', *Philosophy of Science* **53**, pp.584-587.

Forge, J.C. (1993a), 'How Should We Explain Remote Correlations?', *Philosophica* **51**, pp.83-104.

Forge, J.C. (1993b), 'The Modal Interpretation of Quantum Mechanics', *Victorian Centre for History and Philosophy of Science Preprint Series*, 3/93.

Forrest, P. (1988), *Quantum Metaphysics*. Oxford: Oxford University Press.

Hughes, R.I.G. (1989), *The Structure and Interpretation of Quantum Mechanics*. Cambridge Mass.: Harvard University Press.

Redhead, M. (1987), *Incompleteness, Nonlocality, and Realism. A Prolegomenon to the Philosophy of Quantum Mechanics* Oxford: Clarendon Press.

Margenau, H. (1950), *The Nature of Physical Reality*. New York: McGraw-Hill.

Piron, C. (1976), *Foundations of Quantum Mechanics*. New York: Benjamin.

van Fraassen, B.C. (1991), *Quantum Mechanics: An Empiricist View*. Oxford: Oxford University Press

IAN HINCKFUSS

INSTRUMENTAL THEORIES: POSSIBILITIES AND SPACE AND TIME *

Instrumentalism is the belief that good theories about unobservable things are at best merely instruments for generating new truths from ones that are already known. They are not themselves to be taken as true, but merely as useful in the prediction of observable consequences. Realism is the contrary belief that if there is no known falsehood among the logical entailments of some theory conjoined with any set of known propositions, then that theory and all its ontological implications are to be taken seriously. If what seems to be a truth preserving theory implies the existence of atoms and electrons, then we should believe in atoms and electrons, says the realist. If we do not believe in the existential entailments of a theory, then the theory should be discarded.

As far as atoms and electrons are concerned, the realist position seems plausible. But what of numbers? Hartry Field has argued that we need not be realists about numbers, but seems nevertheless to treat the spatial points of geometry with ontological seriousness.[1] Field's position implies that both instrumentalist and realist positions as outlined above are incorrect. If we should regard some theories and their ontological implications realistically and, consistent with this, regard others as mere instruments for generating new truths within a more ontologically economical framework, the question arises: where should we draw the line between the two?

In this essay, nominalism will be assumed to be the theory that theories about only substances and substantial objects, for example, atoms, tables, ghosts, soap and phlogiston should be regarded realistically, and that theories which entail the existence of abstractions, that is, non-substantial objects such as numbers, possibilities, probabilities and properties, should be regarded at best as instrumental. Platonism will be taken to be the belief that at least some theories which entail the existence of abstractions should be taken realistically.

Why should one be a nominalist? A quick answer, though one which will not satisfy many platonists, is similar to the answer given by Benacerraf about mathematical entities; that without substance there is no causation.[2] Without causation, there is no knowledge. Learning about the world is a causal process. Substances but not abstractions are the things involved in causal processes. So what we learn about are substances. Hence abstractions are not objects within the world about which we can learn.

P.J. Riggs (ed.), Natural Kinds, Laws of Nature and Scientific Methodology, 187–209.

The quick answer is useful in showing why the nominalist has those beliefs. It is not very useful for convincing those who have contrary beliefs, for example, that things other than substances can be causal, or that the existence of properties[3] or facts[4] is necessary for causation.

In any case we speak *as if* we know about abstractions. Why should we do this if we do not know of their existence? The instrumentalist-cum-nominalist answer is: because it is logically simplifying to do so and because the practice need not be misleading (despite a couple of millennia of misleading platonist philosophy). This is Field's attitude to numbers.

In this essay, by way of illustrating further avenues for instrumentalism, I shall argue for a particular instrumentalist approach to distances, places, times, events, space, time and space-time, which I shall call *presuppositional eliminativism*. Presuppositional eliminativism will be shown to be immune to arguments contrary to nominalist eliminations of space, time and space-time by John Earman,[5] Graham Nerlich[6] and Hartry Field.[7]

Field's approach to nominalism is via his concept of a *conservative extension*. I shall begin by explaining this idea and I shall then proceed to distinguish between two types of conservative extension, *logically* conservative extensions and *physically* conservative extensions. I shall then attempt to support the case for regarding geometry as a physically conservative extension, and for regarding the denizens of geometry (points, lines, etc.), as fictive.

CONSERVATIVE EXTENSIONS

How do we speak *as if* abstractions exist? In the first place by *presupposing*, though not necessarily *believing*, what Field calls *conservative extensions* of our beliefs about substances and substantial objects. Let me explain that. Consider the following dialogue.

Mary: Suppose Alice goes. Will Jim go?
Grace: Yes. Jim will go.

Although Grace states that Jim will go, she gets committed, in this dialectical context, not to Jim's going but only to his going *if* Alice goes. When she replied, she was talking *as if* Alice were going.

The moral is that what we commit ourselves to when we make a statement that is governed by a supposition is a conditional with the supposition as antecedent and the stated proposition as consequent.

Now suppose that our dialogue is governed by suppositions that do not have to be deliberately introduced, suppositions that are assumed to be in place in any ordinary context, unless they are deliberately withdrawn (as, perhaps, in a philosophical tutorial). Call such suppositions *presuppositions*. Then under normal circumstances we talk *as if* these presuppositions are true. In this way we ordinarily can speak as if abstractions exist and quite rationally disbelieve in the existence of such abstractions at the same time.[8]

What Field calls *conservative extensions*, can be regarded as such presuppositions. What Field calls a *conservative extension* I shall call a *logically conservative extension*, which, for the purposes of this paper will be redefined as follows.

Let L be any language or part thereof with a certain set of allowable predicates. Let T be a set of propositions which require for their expression predicates that fall outside the range of predicates allowable within L, but within the range of some language L_1, which includes L. Say that T is a *logically conservative extension* of L iff, for any proposition **p** expressible in L, and any set S of such propositions, **p** does not follow from $(S \cup T)$ unless it follows from S alone.

Let us call propositions using only the terminology of L, *economical* and the extended terminology of T, *prodigal*.

The interesting cases of conservative theories are those for which the theory T is false, there is an ontological implication, $(\exists x)Fx$, of $(S \cup T)$ that is inexpressible in L, and an economical consequence **p** will follow, if it does, from $(S \cup T)$ *via a proper subset* of the logical rules needed to move from S alone to **p**. In short, a logical simplification is achieved at the price of an artificially inflated ontology and the presupposition of its associated theory. Say that such a theory, T, is a *logically simplifying* conservative extension of L.

For example, allow that L includes sentences featuring the modal operators 'necessarily' and 'possibly', here represented by '$\Box$' and '$\Diamond$' respectively. Allow that the logic for '$\Box$' and '$\Diamond$' is an S5 logic. What is usually but, as we shall see, misleadingly called the formal semantics of S5 allows that:

1. $(\exists w)(w \text{ is a possible world})$

and that for any proposition, **p**, expressed with sentence 'p';

2. $\Box p \equiv (w)(w \text{ is a possible world} \supset \mathbf{p} \text{ is true in } w)$;

and

3. $\Diamond p \equiv (\exists w)(w \text{ is a possible world} \;\&\; \mathbf{p} \text{ is true in } w)$.

The logic S5 is well known to be complete with respect to this 'semantics', that is to say, all arguments expressible in the syntax of S5 and validated by (1) together with appropriate instances of schemas (2) and (3) are derived rules of any natural deduction system which captures S5. Let T be the conjunction of (1) with the instances of (2) and (3). Then another way of saying that S5 is complete with respect to its 'semantics' is: for any proposition **p** expressible in the syntax of S5 and any set S of such propositions, **p** does not follow from $(S \cup T)$ unless it follows from S alone. That is, the formal 'semantics' of S5 comprises a logically conservative extension of the system S5.

That result is quite general. That a logical system is complete with respect to a formal 'semantics' entails that the formal 'semantics' is a logically conservative extension of the logical system.

Let us define some predicate F as being *abstract* iff F is a predicate within some language L_1 used to express some theory T which is a logically simplifying conservative extension of some language L, which does not feature F as a predicate. Define **a** as being an *abstraction* iff F**a**, where F is abstract. Say that L is a *nominalistic* language iff L contains no abstract predicates.

As we have seen, arguments within the economical part of language may be modelled in the ontologically prodigal part of language via extensions which Field calls *bridge laws*, such as (2) and (3) above, which relate prodigal premises T_P and prodigal conclusion $\mathbf{p}_P$ to corresponding economical premises S and economical conclusion **p** respectively. Laws are usually regarded as propositions which are true, but bridge laws may be false, if nominalism is true; which is why so-called formal semantics are not really semantics. Otherwise, the bridge laws would be necessarily true. But propositions which have no existential implications of a certain sort cannot be necessarily equivalent to propositions which do. So 'bridge law' is a misnomer. Nevertheless, for the sake of standard terminology, I shall stick to Field's 'bridge law'. First-order predicate logic is standardly the logic used within the prodigal part of language.

That some theory is a logically simplifying conservative extension of some system of natural deduction is a *prima facie* reason for doubting its veracity. Logical convenience rather than truth can explain the motivation for speaking *as if* the theory is true.

However, that a theory, T_e can be expressed in a language L which is ontologically economical with respect to some logically conservative extension T, is no reason for taking a realist attitude towards the ontology of T_e or any existentially quantified propositions expressible in L. The theory, T_e, in turn, may be an implication of a theory or theories at least some of which are logically conservative extensions of even more economical theories.

PHYSICALLY CONSERVATIVE THEORIES

Instrumental theories need not always be logically conservative extensions of ontologically economical nominalistic theories. Instrumental theories may take us beyond logical validity even if they are truth preserving. Instead of merely providing a model for some laws of logic, as do logically conservative extensions, they may mimic the laws of nature instead. Let us call those extensions which conserve truth without conserving logical validity, *physically* conservative extensions.

How do physically conservative extensions arise? Here's one way. Conjoin some proposition or propositions to a false but logically simplifying conservative theory, T_1, to form a new theory T_2, which is not logically conservative. That is, T_2, unlike T_1, no longer guarantees *logical validity* in the ontologically economical language. It is now logically possible, though not logically necessary, that its conjunction with true economical premises entails some economical falsehoods. If it does, replace T_2 with a modified version T_3, leaving T_1 as one of its conjuncts. If T_3 fails to be truth conserving, replace it in the same way with T_4. Conjecture and experiment can be used, as usual in scientific practice, to approach a truth conserving theory, T_n. Theory T_n will still be false, if only because of its false conjunct T_1.

Why T_n should succeed in conserving truth would demand explanation. The explanation would have to be in terms of the (ontologically economical) true natural laws that are being modelled in the false prodigal ontology. I shall describe someone as being a *presuppositional eliminativist* with respect to at least some of the ontological implications of some theory T, if that person regards T as a logically simplifying conservative extension or a physically conservative extension of an ontologically economical language L in which those ontological implications are not expressible. I shall attempt to illustrate these ideas with geometry cast as a physically conservative extension of propositions about the distribution of matter and the laws governing such distribution.

THEORIES OF PLACES AND TIMES AS INSTRUMENTAL EXTENSIONS OF LANGUAGES ALLOWING SPATIAL AND TEMPORAL MODIFICATION

In this section I wish to argue for the view that spatio-temporal points, lines, areas and volumes are aspects of the richer ontology of physical extensions to language which can describe matter and only matter. If we know that the mouse ran, we can ask *how* it ran (answer: 'Quickly.'), we can ask *where* it ran (answer: 'In the kitchen') and we can ask *when* it ran (answer: 'When the clock struck one.'). Thus we can come to know that the mouse ran quickly in the kitchen when the clock struck one. Superficially at least, our only ontological commitments here are to a mouse, a kitchen and a clock.

Someone may object that we are also ontologically committed to a place, namely, 'in the kitchen' and a time, namely, 'when the clock struck one'. But then why not say that we are also ontologically committed to a manner, namely, 'quickly'? The reason why not is that 'quickly' is not a referring expression. It is not a noun phrase. It is an adverb. But similarly, 'in the kitchen' and 'when the clock struck one' are adverbial. It is true that we would normally equate 'The mouse ran quickly in the kitchen when the clock struck one' with 'The mouse ran in a quick manner, in a place that was in the kitchen, at a time when the clock struck one'. The latter does have ontological commitments to a manner, a place and a time. But does the equation hold because the two are logically equivalent or because the two can be equated via the normally presupposed, though false, bridge laws of a conservative extension?

I suggest that the latter is the case. To make the point, it is necessary to outline what such bridge laws could be. In the case of a spatial adverb, W, modifying a monadic predicate F^1 to describe some object **a**, the ontologically economical version is: $W(F^1)\mathbf{a}$.

Now let **b** refer to the place corresponding to the spatial modifier W. Let P be a predicate meaning 'is a place' and let F^2 be a dyadic predicate meaning 'is F^1 at'. Then the corresponding prodigal proposition is: $(\mathbf{P}\mathbf{b}\ \&\ \mathbf{F}^2\mathbf{ab})$.

Logical simplification results as predicate modifiers are rejected in favour of unmodified predicates. What we have is a schema for an indefinite number of bridge laws of the form:

4. $W(F^1)a \equiv (Pb\ \&\ F^2ab)$

In addition, we could introduce bridge laws for formulae involving 'somewhere' and 'everywhere' (written (SW) and (EW) respectively) as follows:

5. $(SW)(F^1)\mathbf{a} \equiv (\exists x)(Px\ \&\ F^2\mathbf{a}x)$

6. $(EW)(F^1)\mathbf{a} \equiv (x)(Px \supset F^2\mathbf{a}x)$

A supplementary prodigal hypothesis is that there is at least one place, that is,

7. $(\exists x)Px$

The quantifiers '(SW)' and '(EW)' are substitutional simply because their instances are adverbial phrases, not noun phrases. Correct responses to Questions 'Where?' and 'When?', like 'How?' are adverbial phrases such as 'in the kitchen' and 'when the clock struck one'. In contrast, quantifiers like 'somebody' and 'something' may be existential. The proper response to 'Who?' or 'What?' is a name or definite description, and in the case of 'Who?', the name of a person. Instances of existential quantification must be referential and the noun phrase is the medium of reference. When, therefore, we replace 'The mouse ran somewhere' with 'There is a place where the mouse ran' and the question 'Where?' with 'What place?' we are engaging in ontological enrichment.

Valid arguments within a system of natural deduction for spatial modifiers and spatial substitutional quantifiers will mimic the corresponding arguments using existential quantification with reference to places, and instances of (4) to (7) will constitute a conservative extension of such a system.[9]

We could say the same with times and also events as we have done with places, where an event is a place at a time. For example, if 'K' is a spatio-temporal modifier (such as 'in the kitchen at one o'clock') and 'E' is a predicate for 'is an event' and '**k**' names the event corresponding to the spatio-temporal modifier K, then our first bridge law becomes:

8. $$K(F^1)a \equiv (Ek\&{\sim}Ea\&F^2ak)$$

With such an extension, we can speak as if there are many events, but they are all associated with bits of matter. What most of our platonistic friends would like us to provide are the remaining events to fill the gaps, so that we can talk not only of where and when things are or have been or will be, but also of where and when things may be or could have been. Now in providing such an extension, provided that we keep it conservative, we can help ourselves to any abstractions we like, including set theory. Note that people who, like Field, regard spatio-temporal points as substantial must prohibit themselves from such a strategy. However, having already reconciled ourselves to the abstract nature of the prodigal ontology of our extension, we need not feel so constrained. The extension would include the following propositions.

D1. A spatio-temporal point is an event with no events as proper parts.

A1. Any event that is not a spatio-temporal point is the mereological sum of two distinct events.

A2. If two events have no event in common, then there is a continuous set of events whose union overlaps with each.

A3. There is no event that is included in every event.

Allow that such a theory will remain logically conservative with respect to our economical nominalist language. Now, if our theory of space-time is to be anything like the one that physicists talk about, we need it to have geometrical properties. We need a metric to be imposed on the manifiold of spatio-temporal points. Further, the resulting geometry has to be correct as an instrument for generating true geometrical descriptions of physical objects. The basic geometrical property to be assigned to objects is the distance between them. Lengths and breadths of objects can then be defined in terms of the limit of the distance between parts at their boundaries. What is needed is an economical way of saying, for example:

9. The distance from a to b is 5.3 metres.

Note that (9) does not entail the existence of numbers, at least, not without further supposition. The numeral '5.3' is used as a numerical adjective qualifying 'metres', not as a singular term. Some might prefer to say that '5.3' occurs as part of a quantifier, reading (9) as:

10. There are 5.3 metres in the distance from **a** to **b**.

Either way, '5.3' is not being used to refer to a number. However metres and a distance still feature in the ontological entailments of (9) and (10). As an alternative, consider (11):

11. **a** and **b** are 5.3 metres apart.

No distance now, but a question remains about the metres. Clearly '5.3 metres' is acting as an adverb in (11), but it may be argued that 'metres' should still refer to some metres, for all that. It is enormously difficult to say anything without using abstractions, but nevertheless I would argue that 'metres' in (11) is not referential but is modifying '5.3' rather than vice versa. Compare:

12. Earthquake **a** is 5.3 on the Richter scale.

Here 'on the Richter scale' clearly is a modifier for '5.3'. Similarly, although the expression would be slightly unusual, we would not lose any of the intent of (11) if we said instead:

13. **a** and **b** are 5.3 apart on the metre scale.

or better, to rid ourselves of the abstraction 'scale',

14. **a** and **b** are 5.3 apart metre-wise.

Note that (14) is not intended here to be a translation of (13), let alone of (9) and (10). It is intended rather as giving all the economical news intended by assertions of (9), (10) and (13) under the governance of appropriate false but presupposed bridge laws.

Recall that if some prodigal assertion $\mathbf{p}_P$ is governed by some presupposed bridge law,

$(\mathbf{p}_E \equiv \mathbf{p}_P)$,

the speaker commits herself to:

If ($\mathbf{p}_E \equiv \mathbf{p}_P$), $\mathbf{p}_P$

not to $\mathbf{p}_P$.

Given that ($\mathbf{p}_E \equiv \mathbf{p}_P$) is a logically conservative extension, it will be consistent with $\mathbf{p}_E$. Hence, given my analysis of conditionals,[10] which I shall assume here to be correct, if ($\mathbf{p}_E \equiv \mathbf{p}_P$) is a logically conservative extension, 'If ($\mathbf{p}_E \equiv \mathbf{p}_P$), $\mathbf{p}_P$' will be true just in case the facts relevant to the dialogue, together with ($\mathbf{p}_E \equiv \mathbf{p}_P$), entail $\mathbf{p}_P$, and that will be so provided that the relevant facts include $\mathbf{p}_E$. Since $\mathbf{p}_E$ will surely be relevant to any dialogue in which '$\mathbf{p}_P$' is being asserted, it follows that such an assertion will express a truth, namely

If ($\mathbf{p}_E \equiv \mathbf{p}_P$), $\mathbf{p}_P$

just in case $\mathbf{p}_E$ is true.

Here our $\mathbf{p}_E$'s are propositions such as (14) above. In symbols:

15. M(n)**ab**

where M is a measure modifier (e.g. 'apart meter-wise') for the numerical adjective, n (e.g. '5.3'), and **a** and **b** are physical objects as before. The truth of such propositions is determined, if at all, by observation of physical objects, **a, b** and one or more measuring implements. It is not determined by looking where there is nothing to look at. Measuring implements are merely observational prostheses of a sort. Scientific and technological advance may enable us to develop these prostheses and hence refine our observations. But no matter how refined the prostheses the objects observed with them will still be physical objects. Our observational reports may be theory laden, in the sense that what we *think* we observe is a function of the theories we happen to believe at the time; but, for all that, what we *actually* observe is no fiction, even if our theoretical beliefs are false.

The $\mathbf{p}_P$'s corresponding to the economical (15) above will be propositions like (16).

16. $P\mathbf{j_a}\&P\mathbf{j_b}\&O\mathbf{a}\mathbf{j_a}\&O\mathbf{b}\mathbf{j_b}\&M(n)\mathbf{j_a}\mathbf{j_b}$

(Read: $\mathbf{j_a}$ is a place and $\mathbf{j_b}$ is a place and **a** occupies $\mathbf{j_a}$ and **b** occupies $\mathbf{j_b}$, and $\mathbf{j_a}$ and $\mathbf{j_b}$ are n apart metre-wise.)

Spatial and temporal observations, when combined with our bridge laws will give us but a small part of a geometry of space. What is sought for, in the study of physical geometry, is an auxiliary geometrical theory for space or space-time that will entail the abstract spatial and spatio-temporal counterparts of the economical metrical truths. In short, geometry conjoined with the appropriate bridge laws needs only to be *physically* conservative with respect to any ontologically economical language. I have said that the belief that a correct geometry of space or of space-time is a false but physically conservative extension of a nominalist language is an instance of ***presuppositional eliminativism***; 'eliminativist' because such a belief entails that spatial or spatio-temporal regions are non-existent abstractions, and 'presuppositional' because, such a belief is consistent with talking *as if* the ontological implications of such a geometry were true, given that the geometry and its appropriate bridge laws are *presupposed*.

Despite the plausibility of a presuppositional eliminativist account of geometry, many philosophers have argued in favour of an ontology which includes points, lines, volumes, times and events. Let us now examine some of those arguments in the light of the possibility of presuppositional eliminativism.

ARGUMENTS FOR THE REALITY OF SPACE OR OF SPACE-TIME

Argument 1. Theories of absolute space and space-time in several different senses of 'absolute' are meaningful, possible and consistent with both Newtonian dynamics and Relativity theory (John Earman).[11]

Response. Earman's article is a defence of absolutism in theories of space-time against 'relationalist' attacks which purport to demonstrate the inconsistency or meaninglessness of absolutist theories. However, with one exception, in all the senses of 'absolute' which Earman is willing to allow as plausible for space or space-time, presuppositional eliminativists can also allow descriptions of absolute space or space-time in theories which they regard as plausible possibilities as conservative extensions of their nominalist base language. Further, for any such theory to be a conservative extension would require that it be both consistent and meaningful. If it were not semantically meaningful, it is hard to know why it would count as a theory at all. In any case, it would be useless as an extension to any theory, nominalistic or otherwise. Again, if an extension were inconsistent, it could not be conservative with respect to any nominalist base language.

Some positivists have suggested that, though semantically meaningful, such theories may not be *empirically* meaningful, meaning that there may be no way that the theory could be tested for truth or falsity. There has been difficulty in making sense of the modalised predicate 'could be tested',[12] but in any case, the presuppositional eliminativist will be interested only in the theory's instrumental success. Whether *all* its implications, including the existence of spatial or spatio-temporal points, could be verified or falsified would be irrelevant.

The exception referred to above was Earman's first sense of 'absolute' meaning 'objectively real, existing "without the mind" and "distinct from that which is perceived by the senses" ', the quotations being from Bishop Berkeley.[13] There is no quarrel here with Earman's criticism's of Berkeley's anti-Newtonian arguments.[14] Berkeley's arguments can be interpreted as a case against the possible instrumental success of Newton's theories of space. Any theory that is not instrumentally successful is either false, because it yields the wrong predictions, or is useless because it yields no novel predictions either by itself or when conjoined with other theories. A presuppositional eliminativist could allow, if it were so, that Newton's theories were instrumentally successful but still allow that their spatial ontological entailments were false.

Argument 2. Space is either orientable or it is not. Neither disjunct of that disjunction is analysable as a statement concerning material objects and only material objects. Hence Space is absolute. The orientability or otherwise of Space could explain why objects that are asymmetrical in all dimensions cannot or can, as the case may be, be made congruent with their mirror image simply by rigid transportations and rotations. Hence it is plausible to think of space as a feature of the world, that is, as existent.[15]

Response. The above argument is my attempt to precis the argument with which Nerlich has renewed respect for a similar argument put forward by Immanuel Kant.[16] The presuppositional eliminativist would nevertheless remain unconvinced.

Since the opening proposition of the argument refers to Space meant as the mereological sum of all places, the presuppositional eliminativist might reject this as false and question begging to boot. But that would be too swift. The presuppositional eliminativist is used to talking as if various falsehoods were true. So she might take the opening sentence to be committing her rather to the proposition that *if* Space existed, it would be either orientable or not, to which she may well concur. Further, she would be correct in being convinced by Nerlich's more detailed arguments that 'Space is orientable' is not to be analysed in terms of the doings of material objects and that therefore Space is absolute in that sense. She could agree also that the orientability or otherwise of Space could explain (again given the supposition that Space existed) why objects cannot or can be converted to their mirror images by rigid transport and rotation. But why should she leap from that to believing that Space is *in fact* a feature of the world?

The answer would presumably be that it is reasonable to believe the best possible explanation of any phenomenon. However, the presuppositional eliminativist will not see it as a good possible explanation, let alone the best one. Believing, as she does, that Space is a construct out of a fictive theory of places introduced for the purposes of logical simplicity, she will have a fictive view of Space and will regard theories of Space as no more explanatory *in fact* than theories of Santa Claus or the tooth fairy.

Further, the presuppositional eliminativist can agree that the fictive theory of Space has many uses, not only within topology and geometry, but also within the many field theories. But although she can agree that it is good or fruitful for providing a model onto which may be built many different instrumentally successful theories, she will not be tempted into regarding this proliferation as a rationale for the truth of the theory, knowing its origins as she does. Although it may be true, as Michael Friedman[17] and others have said, that a 'theoretical structure that plays an explanatory role in many diverse areas picks up confirmation from all those areas', it does not follow that a theoretical structure that plays an instrumentally useful role in many diverse areas is automatically one which plays an explanatory role in all those areas.

Kant, by the way, could be interpreted as a presuppositional eliminativist in so far as he thought that, though space and time were unreal, all our thoughts had spatial and temporal implications, that is, we are obliged to think *as if* both space and time existed. In other words, it is human nature to presuppose the existence of space and time.

Argument 3. Unlike our knowledge of the mathematical structure of real numbers, which, if anything, is *a priori*, the structure of physical space is an empirical matter (Hartry Field).[18]

Response. Field intended this argument primarily to point out what he took to be an important ideological distinction between real numbers and space-time points. However, just as the ideology of 'number' can be introduced via logically conservative extensions of ontologically economical logical systems, so the ideology of 'place' and 'event' can be introduced by logically conservative extensions of economical logical systems. Further, just as these conservative extensions could themselves be extended to produce interesting and useful geometries that, though no longer logically conservative, would be nevertheless *physically* conservative, that is, truth preserving by virtue of the way the world works, so, as Field himself tells us, the conservative extensions of number theory can be extended to platonically expressed false, though instrumentally correct, counterparts of the laws of physics. It is empirical research in both cases that informs us of the instrumental correctness of the platonically expressed counterparts. Indeed, those counterparts of physical geometry which utilise number concepts are just a case in point. Field says that mathematics was developed to deal with physical space. Ignore the assumption here that there is such a thing as physical space.[19] It remains the case that such developments to which he takes himself as referring (namely geometry as we normally know it) would have been at best empirically demonstrable as instrumentally correct extensions of the union of a nominalist language with the language of number theory. It is just as valid to regard mathematised geometry in that way as to regard it as the result of a number-theoretical extension of the union of a nominalist base language with a geometrical language.

Argument 4. In contrast to substantivalism, i.e., the belief that space-time points and regions exist as entities in their own right, and are 'full-fledged causal agents',[20] are reductive relationalism and eliminative relationalism. (This terminology will be explained below.) Reductive relationalism entails the existence of sets and so is unavailable to nominalists.

Both relationalist theories have difficulty in handling the notion of a field. Substantivalism is both nominalistic and capable of dealing with field theory. Relationists 'would have to predict and explain the behaviour of matter in terms only of that matter and other matter (e.g., the matter that a substantivalist might intuitively think of as 'generating' the relevant aspects of the field)'.[21] Hence such theories would require that there be action at a distance.

Response. Field would not see this as a knock-down case for substantivalism. His intention is merely to show that substantivalism has certain advantages over what he calls relationalism. However, though presuppositional eliminativism is anti-substantivalist, and in that sense relationalist, it does not entail either of the relationalist positions Field speaks of. Further, it denies the existence of any abstractions at all, including sets and numbers and so is fully nominalistic. I shall deal with the problem of action at a distance as a separate argument (Argument 7). Meanwhile, let us look at the difference between presuppositional eliminativism and the sorts of relationalism that Field has in mind and the problems raised by the notion of a 'field'.

Reductive relationalism is the view that 'points and regions of space-time are some sort of set-theoretic construction out of physical objects and their parts' says Field,[22] by which I take it he means that reductive relationalists would have it that points and regions of space-time are definable in terms of sets and physical objects. Presuppositional eliminativism, however, rejects any such semantic equivalence.

Eliminative relationalism, according to Field,[23] is the belief that 'it is illegitimate to quantify over points and regions of space-time at all.' Presuppositional eliminativism allows such quantification, provided only that it is done under the governance of appropriate presuppositions to eliminate what would otherwise be the ontological commitments of such quantification. 'If p, (∃x)Fx' does not entail '(∃x)Fx', and given that the presuppositions of abstraction generating extensions are false, according to presuppositional eliminativism, there is no chance of producing such an entailment with *modus ponens* using true premises.

On one account, a field is a function which assigns a physical quantity to each spatio-temporal point. On such an account, then, the existence of fields entails the existence of space-time points. But fields, being functions, are abstractions. So a nominalist, such as Field, cannot rationally believe in fields either. What he can do, and what he does, however, is to believe field *theories* which he takes as being theories which entail propositions such as $\varphi\mathbf{k}$ where φ is a predicate assigning a causal property to a space-time region or space-time point, **k**.

Substantivalism is consistent with such theories, and clearly the two relationalist theories above are not. Neither is presuppositional eliminativism. However, presuppositional eliminativism can treat field theories, including the more platonically expressed varieties, as extensions of the space-time theories already established as extensions of propositions about material objects for geometrical purposes. Such extensions again may or may not be physically conservative, and were they found not to be so, they could be dropped as instrumentally unsuitable suppositions. Unlike reductive and eliminativist relationalism, presuppositional eliminativism could therefore cater for field theories, without any ontological commitment to either fields or space-time points. Hence, even if field theories which ascribe properties to regions of space-time are important to physicists, this in no way tells against presuppositional eliminativism.

Field defines relationalism in his *Realism, Mathematics and Modality* as the belief that 'the physical world contains spatio-temporal aggregates of matter' ... which 'are interrelated in various ways by various geometric (and also non-geometric) relations, but the physical world does not contain a space-time over and above these aggregates of matter'.[24] Presuppositional eliminativism clearly is a species of relationalism in that sense.

Presuppositional eliminativism is also a species of what Field calls *predicate monadicism*, the idea of reducing one's ontological commitment to places and events by regarding predicates such as 'occurs at time t' as a monadic predicate.[25] Recall that our presuppositional eliminativist treats adverbial modifiers such as 'at time **t**' for what they are — adverbial modifiers which do not change the addicity of the predicate modified. The presuppositional eliminativist insists, correctly, that '$J(F^1)\mathbf{a}$', is *not* logically equivalent to '$F^2\mathbf{aj}$', where 'F^2' is dyadic and **j** is some place corresponding to the spatial modifier, 'J'. The substantivalist may consistently (but falsely, according to presuppositional eliminativism) assert in presupposition-free dialectical mode that the two are materially equivalent. The presuppositional eliminativist will not be so bold. She will be content with merely *presupposing* that material equivalence. This brings us to Arguments 5 and 6 against presuppositional eliminativism in the form of arguments which Field raises against predicative monadicism.

Argument 5. It 'is unclear what the monadicist analogue is of substantivalist assertions that quantify over space-time points or other space-time regions'.[26]

Response. The presuppositional eliminativist can offer a very clear analogue of such assertions. The substantivalists assertions are *presupposition free*. The corresponding assertions for the presuppositional eliminativist use exactly the same sentences, but she regards her assertions as *governed* by presupposed bridge laws (such as those that relate modified monadic predicative constructions to dyadic predicative constructions) plus any auxiliary theory (such as mereology) about the denizens of such prodigal spatio-temporal ontology as has been introduced by the bridge laws. The effect is that the presuppositional eliminativist effectively asserts a modest conditional, where the substantivalist boldly and unconditionally asserts the consequent of that conditional.

Argument 6. Predicate monadicism looks unattractive because 'to accept a monadicist theory would be to accept a theory with an infinite (indeed, uncountably infinite) stock of semantically primitive predicates — one corresponding to each of the space-time points or regions recognised by the substantivalist'.[27]

Response. If presuppositional eliminativism needed such a stock of predicates, it would not only be unattractive but an impossibility, since no language can have an uncountable number of predicates. However, by the same token, a language can have at most only countably infinite names and the presuppositional eliminativist can use her bridge law to yield a position predicate 'occurs at **k**' for any name '**k**' that the substantivalist uses. Although she cannot name them all, it is the substantivalist, if anyone, who will want to assert unconditionally that the number of spatial or spatio-temporal regions is uncountably infinite. The presuppositional eliminativist has no need to do that.

To suggest, however, that the presuppositionalist derives her spatial or spatio-temporal predicates from substantivalist names of spatio-temporal regions is the reverse of what actually happens in practice. For a radar operator, it will suffice to know that the aeroplane is at range 50 kilometres, bearing 80° north and elevation 10°. She will not be concerned as to whether there is a substantial spatial region there as well as the aircraft. (If this thought did occur to her, she might be concerned for the safety of the occupants of the aircraft.) Indeed, not even the substantivalist will believe that the name of a physically significant substantial spatial point is obtainable from that spatial description of the aircraft; for that description is indexical. It is a description that is relative to a particular radar. That is not to say that we cannot, via the operation of false platonic bridge laws, talk of the place where the aeroplane is to be found as being at a range of 50 kilometres, etc., but no such place would be regarded as a substantial object or as one of the regions of an absolute and real space even by those who are inclined to believe in real space. Practical physicists will always have a need for spatial and spatio-temporal predicates and predicate modifiers and a denumerable infinity of such modifiers is more than they will ever need. Note further that the predicate modifiers available to the physicist and the presuppositional eliminativist are not, and need not be, semantically primitive as Field suggests they must be. In practice, they are generated from a finite range of nouns, adjectives, adverbs and prepositions, together with denumerably infinite numerical adjectives, each of which is constructed as a finite sequence of symbols chosen (in the case of a decimal radix) from ten possibilities.

As an aside, note in passing that it would be a mistake to suppose that someone who believed that an aeroplane was at a range of 50 kilometres was ontologically committed to a range of 50 kilometres. The mere presence of a noun or noun phrase does not indicate any such commitment. Consider for example: 'There is no such thing as a flying horse.' Even where there is no overt denial of ontological commitment, the formation of adjectival and adverbial phrases from noun phrases and prepositions may suffice to eliminate any ontological commitment of the corresponding noun phrase. Consider for example: 'I was dreaming of a flying horse.' There is no question as to whether the speaker is related to an actual flying horse by some relation of 'was dreaming of '. The speaker need not believe in the existence of any such horse in order to believe that she has been dreaming of one. For this reason it makes sense to parse 'of a flying horse' as an adverbial modifier of the monadic 'was dreaming' resulting in the equally monadic 'was dreaming of a flying horse'.

Argument 7. A physical theory which recognises the existence of neither spatio-temporal regions nor substantial fields is committed to replacing field theories by action-at-a-distance theories.

Response. Argument 7 assumes that causal propagation via a field takes place over non-zero distances. However, in modern physics, such propagation takes place at the speed of light and spatio-temporal distances along such 'paths' are zero. Hence if the field, and the gravitons and photons with it were merely presupposed fictions, there would be no causal spatio-temporal gap to bridge and hence no action at a spatio-temporal distance to worry about.

If relativity one day became unpopular with physicists, and spatio-temporal distances along the causal paths of fields were no longer regarded as of zero magnitude, it might turn out that action-at-a-distance theories would once again be popular. (It is only a few decades since physics students were being taught that gravitational forces were forces that acted at a spatial distance.)

Alternatively, people might regard non-zero spatio-temporal distances of electromagnetic and gravitational interaction as a further reason for taking the existence of fields, gravitons and photons seriously, as many do today. Field mentions Malament and Putnam as cases in point.[28] These authors point out that if one takes fields themselves as substantial, the motivation from field theory for treating space-time as substantial vanishes, easing the way for relationalism. Similarly, it would ease the way for presuppositional eliminativism, if that were necessary.

Field believes that this approach to field theory trivialises relationalism. He says that on any 'interesting version of relationalism, fields are just as much a fiction as space-time.' There are two possibilities to discuss here. The first is that fields should be treated as substantial objects. The second is that fields should be treated as abstractions. If fields should be treated as substantial, then if presuppositional eliminativism were correct, spatial and spatio-temporal regions would still be insubstantial abstractions. If the world were a plenum, spatio-temporal regions would still be distinct from the matter that occupied them. In the presupposition-free discourse of the presuppositional eliminativist, there would be no spatio-temporal regions, but matter would be everywhere. With the spatio-temporal presuppositions operative, statements to the effect that spatio-temporal regions exist would be in order. However, matter would be said to occupy them and not vice-versa. The two would not be identical. A piece of matter is not to be identified with the place where it is to be found. So substantivalism would still be false.[29]

Now if fields are to be treated as abstractions, the presuppositional eliminativist can eliminate fields in exactly the way that she eliminates space-time. Any instrumentally successful theory that entails that materially empty spatio-temporal regions or points have physical properties can be regarded as a presupposed fiction. It will be instrumentally successful only if it predicts the properties of pieces of matter. Matter is, after all, what we observe and it is on such observations that we judge a theory to be instrumentally successful. Theories about fields could be given exactly the same treatment by presuppositional eliminativists as they give to geometry. Assertions which ascribe seemingly causal properties to empty spatio-temporal regions could be regarded as being governed by presuppositions, which reduce the commitment of such an assertion to a harmless conditional. Corresponding assertions for materially occupied regions would be replaceable via the appropriate bridge laws giving those causal properties to pieces of matter rather than pieces of space-time.

Argument 8. Substantivalism has no need for conditionals; but Relationalism has a need for conditionals, for instance to define what it is for the distance between z and u to be twice the distance between x and y. However,

(i) conditionals are vague;

(ii) if distance relations are defined with conditionals, 'then situations that differ in their distance relations ... must differ in some' non-conditional 'respects as well'. If 'a relationalist could find a way to grant this, he or she would have no need of ' conditionals; and

(iii) 'no theory about' conditionally 'defined relations will be forthcoming, unless those relations can also be defined' non-conditionally.[30]

Responses.

(i) The alleged vagueness of conditionals.

Clearly, presuppositional eliminativism has a need for conditionals. I have used 'conditionals' where Field has used 'counterfactuals' which I take to be the standard misnomer for sentences containing an 'if' clause or the linguistic equivalent. There is no syntactic or semantic type which can properly be said to be counterfactual. A counterfactual conditional is a conditional whose antecedent is false. If we are talking about conditionals which may or may not be used 'in the formulation of physical theory' we had better allow that on occasion the antecedent of such a conditional may be true.

It is a dogma widespread among philosophers of science that conditionals are vague or that they have strange ontological implications concerning 'possible worlds' and that therefore their use should be eschewed in the formulation of physical theory. This animadversion is often extended to any sort of modalised proposition.

However, this puzzlement about and bias against modal words including conditionals does not seem to be shared by the public at large, let alone physicists in the development of their theories. Just as one example, take section 2 of Chapter 1 of Frank's *Introduction to Electricity and Optics*, entitled, significantly enough for our present purposes, 'Conservative and Nonconservative Force Fields; Potential Energy'.[31] Here Frank gets into the nitty-gritty after a small historical sketch and a page which describes the difference between a vector and a scalar field. Section 2 covers slightly less than three pages of text and contains forty-eight sentences. Of these, eight contain the word 'if', one of them twice. Another uses 'since'. Ten contain one of the modalities, 'can', 'cannot', 'must', 'it is necessary that', or 'the picture of' meaning 'the possibility of'. Seven of the sentences introduce suppositions and are therefore to be counted as the antecedents of conditionals. Thirteen of the sentences are clearly governed by these suppositions, and so act as the consequents of conditionals whose antecedents are one of the seven suppositions. (A supposition may govern more than one sentence.) There are at most twelve sentences out of the forty-eight that are *not* modalised or involved in conditionals. Further, two of the three definitions in the section contain the word 'if'. A quick glance at other pages of the text reveals a similar proliferation of modalities and conditionals, and the case is similar for other physics texts. By the way, there is nothing vague about the section from Frank's book just mentioned. It is the acme of clarity and precision.

It is one thing to be able to do something well and another to understand how one does it. These two get confused in semiotic discourse, for the thing one does well is to understand the language one habitually uses. What semioticians try to understand and what we all find very hard to understand is how we understand our language. I conjecture that many philosophers are wont to confuse the understanding of some aspect of our language with the understanding of our understanding of that aspect of our language. If we did not understand conditionals well we would be wise to leave them out of our discourse. But our ignorance about how we understand conditionals would be no reason to bar their usage.

In any case, a clear and fairly simple account of the pragmatics and semantics of conditionals, which accords well with ordinary language, is now to hand.[32] The idea, briefly, is as follows. At any stage of any dialogue the dialogue will be concerned with a certain subject-matter (electricity or fields for example). Call the set of truths concerning the subject matter, the relevant facts. In general, the set of relevant facts will be a proper subset of the set of truths. It will not necessarily be closed under entailment. Consider all the consistent conjunctions (if any) of propositions including the antecedent of the conditional, any suppositions governing the conditional and a maximal number of members of the relevant facts. The conditional expresses a truth iff *either* all such conjunctions entails the consequent *or* (in case there are no such conjunctions) the antecedent conjoined with the suppositions alone entail the consequent. This is a development of the *ceteris paribus* analysis by Angelica Kratser,[33] which in turn is a development of Frank Ramsey's analysis in 'General Propositions and Causality'.[34]

(ii) and (iii) Defining Distance Relationships.

Field suggests that what he calls a 'strong moderateness' condition must be satisfied if ontological commitment to numbers is to be avoided.[35] The condition is that:

(a) that it is possible to state the laws of physics in such a way that they do not involve magnitude relations between physical things and numbers, but only involve relations between physical things alone; and

(b) that any 'formulation of the laws of physics in terms of magnitude relations between physical things and numbers can be derived from the more basic formulation of these laws' (which do not involve such relations) 'via the representation theorems for the magnitude relations'.

He claims that the substantivalist has no difficulty in meeting these conditions, but that it is doubtful that a relationalist theory could do so. It turns out that this is so because it is possible for the substantivalist to define the magnitude relations between different distances in terms of two basic relations 'is congruent with' and 'is between', and reference to appropriate spatial points that are always everywhere in between any two other spatial points. The relationalist, on the other hand cannot, in general, refer to such substantial spatial points.

Again, the presuppositional eliminativist can match the substantivalist when it comes to definitions in terms of spatial points. The mere fact that one is talking of fictions does not prevent one from defining the terms used to describe them. But that does little to satisfy the spirit of Field's program which, if I understand it correctly, is to be able to explain facts like why some distance, from x to y, say, is n times the distance from z to w, in terms of relations between substantial objects. For the substantivalist, according to Field, the answer is fairly readily to hand. Where n is 2, for example, it is by virtue of the existence of a *real* substantial spatial point, u say, between x and y such that the distance from z to w is congruent with both the distance from x to u and the distance from u to y. Relationalists and in particular, the presuppositional eliminativist, cannot, in general, with their limited ontology, point to such real intermediaries as the substantial point u.

Is Field's program really necessary? Where are the references to numbers in Newton's law of gravitational attraction, for example? The law could be expressed as:

The mutual gravitational force between two point masses is proportional to the product of the two masses and inversely proportional to the square of the distance between them.

This version talks of masses and distances, but not of numbers, unless one takes products and squares as numbers. But it is not at all clear that one should do that. The square of a distance could be taken to be an area, not a number. Again, if it were a number, what number would it be? That would depend on the unit of length that was adopted. So the square of the distance cannot be referring to any particular number. It is possible, therefore, to think of the law as expressing a relationship between certain quantities rather than between certain numbers.

So although we may agree that this formulation of Newton's gravity law relates certain quantities or magnitudes to one another, we may not agree that, by itself and without further extensions as extra premises, it entails the existence of numbers. To support the proposition that there is such an entailment, one must show how 'There are numbers' is derivable from the law itself, using only valid arguments, and without assuming the logical necessity of any ontology expanding bridge laws.

Is Field's program desirable? It would be desirable if it gave us an understanding of why different distances have the magnitude relations that they do, for example why the distance from x to y is twice the distance from z to w. But it is hard to see how Field's program of reductions via his 'representation' theorems does that. The theorems define what it is for a magnitude ratio to have value n in terms of certain primitives, including one which tells us what it is for a magnitude ratio to be unity. In the case of distance comparison, the primitive in question is 'is congruent with'. Now, necessarily, if **a** is congruent with **b**, **a** shares all of **b**'s distance measurements. So to understand congruence we have to understand what it is for two distances to be the *same* magnitude, that is, for their magnitude ratio to be unity. But there seems no reason to believe that this should in any way be more comprehensible than the possibility of one distance having twice the value of some other distance. The question 'By virtue of *what* does this magnitude ratio obtain?' presents the same poser regardless of whether that ratio is one or fifty-one.

Distance ratios are usually defined in common practice and by physicists in terms of possible transport and superposition of rigid rods — what would happen *if* the rods coincided when superimposed and then were subsequently transported etc.. Field is correct to think that such conditionals should have a nonconditional basis (what I have called the relevant facts). He is correct also in saying that it would be interesting to learn these relevant facts. The same applies to theoretical support for the conditionals to which the presuppositional eliminativist is committed in taking geometry to be at most instrumentally correct. The search for such theories is important to our understanding of the way the world works. Field would describe such theories as *intrinsically* explaining what the conditionals of the presuppositional eliminativist would explain only *extrinsically*. As Field correctly says 'one wants to be able to explain the behaviour of the physical system *in terms of the intrinsic features of that system*, without invoking extrinsic entities ... whose properties are irrelevant to the behaviour of the system being explained'.[36]

Now it is just such theories that Field demands of relationalists and, in general, realists demand of instrumentalists before they are prepared to give up their realism. But this attitude is wrong, I would claim. It is wrong firstly because it lumbers the realist with false beliefs. It is wrong secondly because it dissuades the realist from the search for further physical laws that govern the behaviour of matter. It amounts to a pessimistic and conservative constraint on what are believed to be the limits of human understanding and hence on future theoretical research in physics.

Argument 9. If we are going to be instrumentalist with respect to space-time regions and fields, why not be instrumentalist with respect to atoms and molecules or any other theoretical entities?

After all, there are those who are substantivalist with respect to space-time and those who are substantivalist with respect to fields. So if you are going to withdraw substantial space-time and substantial fields from your ontological commitments, it cannot be because they are not substantial. Where, if anywhere, should we draw the line between an ideology and an ontology that is instrumental and an ideology and an ontology that is real?

Response. Substantiality, the supposed possession of physical properties that are causally efficacious, is necessary for physical reality, but it is not sufficient. Santa Claus and the tooth fairy are supposed to have physical properties that are causally efficacious, but for all that, they do not feature in the ontology of even a small proportion of adults. Stories of Santa Claus and the tooth fairy are invented for the amusement of children. Part of the children's interest comes from the explanatory role that is given to these stories about the appearance of gifts and the replacement of a tooth by a coin. But adults have alternative explanations for these phenomena. They *know* that it was themselves who organised the appearances and disappearances. It is not a matter of inference to the best explanation. The correct explanation is known from the start. The ideology of 'Santa Claus' and 'tooth fairy' was not invented for providing a correct explanation of certain phenomena and that is why adults do not believe the corresponding stories.

Similarly, if one can see that the introduction of certain predicates involves motivations other than correct explanation of phenomena, for example, to achieve a simpler logical structure for the sentences one uses in one's theoretical deductions, then in presupposition free contexts such as philosophical discussions of ontological matters, one has a good reason not to take seriously the existence of items satisfying such predicates.

The predicates 'atom', 'molecule', 'electron' and 'quark' were introduced neither for the amusement of children nor for any logical or other heuristic convenience. They were introduced in an endeavour to provide correct explanations of certain phenomena and for that purpose alone. For all that, it is possible that some such predicates fail to describe anything that is real, and that we could be mistaken in believing that they do. But such a mistake will not involve being confused by the pragmatics of our own linguistic habits.

CONCLUSION

In this essay I have argued that instrumentalism for particular physical theories can be a viable option to realism. I have taken the instrumentalism of geometry as a case study, and in doing so, I have critically examined and found wanting a number of arguments commonly used among philosophers of science for realism and against instrumentalism for particular theories of physics.

University of Queensland, St. Lucia, Queensland

NOTES

* This paper was presented to the annual conference of the Australasian Association for the History, Philosophy and Social Studies of Science in Brisbane, Queensland in 1994. Thanks are due to the University of Queensland for providing the study leave during which a first draft of this paper was written, and also to colleagues at the University of Queensland, to Jack Smart, Hugh Mellor, Jeremy Butterfield, Peter Lipton and Jonathan Harrison for useful discussion of that draft.

[1] Field (1980).
[2] Benacerraf (1973).
[3] See, for example, Armstrong (1978).
[4] See, for example, Mellor (1991).
[5] Earman (1970).
[6] Nerlich (1976).
[7] Field (1980), Ch.4; (1989), Ch.6.
[8] For further detail, see Hinckfuss (1993).
[9] The idea of a logic for spatial modifiers matches the idea of a logic for temporal modifiers such as that proposed by Butterfield and Stirling (1987).
[10] Hinckfuss (1990).
[11] Earman (1970).
[12] Hempel (1952).
[13] Berkeley (1951), Vol. 2, p.93.
[14] Earman (1970), pp.306-10.
[15] Nerlich (1973) and (1976), Chapter 2.

16 Kant (1968), p.37.
17 Friedman (1983), p.243.
18 Cf. Field (1980), p.31.
19 Field (1980), p.33.
20 Field (1989), p.34 & p.114, note 23.
21 *ibid.*, pp.181-82.
22 Field (1980), p.34.
23 *ibid.*
24 Field (1989), p.171.
25 *ibid.*, pp.175-76.
26 *ibid.*, p.176.
27 *ibid.*
28 Field (1989), p.182, n.19; Malament (1982), n.11; Putnam (1981), p.73.
29 Cf. Friedman (1983), pp.221-23.
30 Cf. Field (1989), p.211.
31 Frank (1940).
32 In: Hinckfuss (1990).
33 Kratser (1981).
34 Ramsey (1929) in Mellor (ed.) (1978).
35 Field (1989), pp.188, 189.
36 *ibid.*, p.193.

REFERENCES

Armstrong, D. M., *Universals and Scientific Realism*, Cambridge: Cambridge University Press, 1978.
Benacerraf, P., 'Mathematical Truth', *Journal of Philosophy* **LXX**, 19, (1973): 661-679.
Butterfield, J. and Stirling, C., 'Predicate Modifiers in Tense Logic', *Logique et Analyse* **117-118** (1987): 31-50.
Earman, J. 'Who's Afraid of Absolute Space?', *Australasian Journal of Philosophy* **48** (1970): 287-319.
Field, H., *Science Without Numbers*, Oxford, Basil Blackwell, 1980.
—, *Realism, Mathematics and Modality*, Oxford, Basil Blackwell, 1989.
Frank, N.H., *Introduction to Electricity and Optics*, New York, McGraw-Hill, 1940.
Friedman, M., *Foundations of Space-Time Theories*, Princeton University Press, 1983.
Hinckfuss, I., 'Suppositions, Presuppositions and Ontology', *Canadian Journal of Philosophy* **23** (1993): 593-616.
—, 'Relevant Facts and Suppositions: A New Analysis of Conditionals', *Logique et Analyse* **131-132** (1990): 215-41.
Hempel, C.G., 'Problems and Changes in the Empiricist Criterion of Meaning' in Linsky, L., (ed.) *Semantics and the Philosophy of Language*, University of Illinois Press, 1952, pp.163-85.
Hughes, G.E. and Cresswell, M.J., *An Introduction to Modal Logic*, London, Methuen, 1968.
Kratzer, A., 'Partition and Revision: The Semantics of Counterfactuals', *Journal of Philosophical Logic* **10** (1981): 201-16.
Malament, D., 'Review of Field (1980)', *Journal of Philosophy* **79** (1982): 523-34.
Mellor, D.H., (ed.), *Foundations*, London, Routledge and Kegan Paul, 1978.
—, *Matters of Metaphysics*, Cambridge: Cambridge University Press, 1991.
Nerlich, G., 'Hands, Knees and Absolute Space', *Journal of Philosophy* **70** (1973): 337-51.
—, *The Shape of Space*, Cambridge: Cambridge University Press, 1976.
Putnam, H., *Reason, Truth and History*, Cambridge: Cambridge University Press, 1981.

PETER J. RIGGS

SPACETIME OR QUANTUM PARTICLES: THE ONTOLOGY OF QUANTUM GRAVITY? *

What is called a law of nature tends to depend on the historical circumstances in which the law was discovered or inferred. In the mathematical sciences generally, many laws can be expressed as an equation (or set of equations) which hold within a specific sphere of applicability. Our best confirmed theory of gravity is general relativity which is applicable to large scale phenomena. Quantum theory, on the other hand, is the most empirically successful theory at microscopic scales. However, we do not tend to speak of individual laws of general relativity or of quantum theory in the same way as we speak of Newton's Laws or the Laws of Phenomological Thermodynamics.

What results when theories/laws that have been developed in one sphere of applicability are applied to another (very different) sphere? The domains of quantum theory and general relativity overlap in situations where quantum mechanical effects cannot be ignored, such as might exist near (or inside) a black hole, or in the early universe. In order to deal with this overlap of theoretical domains, there has been a tendency to apply the rules of quantum field theory to the classical gravitational field equations and without much regard for the implications of the whole enterprise. These implications will be the primary concern of this essay and not the status of physical law in general relativity and/or quantum theory.[1]

Since its inception by Einstein in 1915, the general theory of relativity has been interpreted in realist terms. General relativity describes a classical (i.e., not quantum) gravitational field by means of its field equations, viz.:

$$G_{\mu\nu} = R_{\mu\nu} - \tfrac{1}{2}\, g_{\mu\nu}\, R = -\,\kappa\, T_{\mu\nu} \qquad (1)$$

where $G_{\mu\nu}$ is the Einstein tensor, $R_{\mu\nu}$ is the Ricci tensor, $g_{\mu\nu}$ is the metric tensor, R is the Ricci scalar, $T_{\mu\nu}$ is the stress-energy-momentum tensor and κ is Einstein's constant. Classical general relativity is a theory about the *geometry of spacetime*.[2] The presence of matter (represented by $T_{\mu\nu}$) curves spacetime and the motion of matter is determined (in part) by the curvature of the spacetime (given by $g_{\mu\nu}$) through which the matter moves.

Quantum field theory, on the other hand, does not lend itself to a strongly realist interpretation, unlike general relativity. Despite the fact that quantum field theory does give accurate predictions for the interactions of 'elementary'

P.J. Riggs (ed.), Natural Kinds, Laws of Nature and Scientific Methodology, 211–226.

particles, it is not difficult to construct convincing arguments to the effect that much of quantum field theory is a theoretical fiction.[3] This makes the realist assessment of quantum field theory far more problematic.[4] However, due to the large amount of experimental evidence that has been collected, most quantum particles are accepted by physicists as real entities and are classified into families according to their physical characteristics, e.g. mass, charge, spin, etc.

Researchers in quantum gravity, for the most part, can be divided into two broad groups which we shall denote as 'relativists' and 'particle theorists'. The former adhere to the geometric view of gravity, whilst the latter favour approaches in which gravity is a force mediated by one (or more) quantum particles.[5] Our best scientific theories when interpreted realistically ought to inform us about the sort of entities we should have ontological commitments to. Indeed, as Nerlich has remarked, "scientific realism ... what is that if not ontology?" [6]. Why do the particle theorists take such a different line to the relativists in regard to gravity when classical general relativity is so well confirmed? The primary reasons would seem to be as follows:[7]

(I) The right-hand side of the field equations (1) describes matter sources, the behaviour of which is governed by quantum theory. The left-hand side of the field equations describes gravitation as a classical field. If the right-hand side represents quantized matter then the field equations as they stand are *inconsistent*.

(II) The Uncertainty Principle implies that the gravitational field cannot be measured to arbitrary accuracy. The measured strength can only be given as an average over a spacetime region and not at individual spacetime points.

(III) The unification of all fundamental interactions under a single theoretical edifice has been a long-standing goal in physics. Recent successes (e.g. the Electroweak theory) have provided additional impetus for further unification endeavours (including gravity).

(IV) The removal of physical singularities. Classical general relativity predicts the existence of spacetime singularities. These are considered a blemish on spacetime models because all physical laws are assumed to break down at a singular point.

Although there are sound grounds for (I)–(IV), it should be noted that none of these reasons are *logically* compelling.

The dichotomy between geometric and particle pictures of gravity is evident especially in the case of gravitational waves. Consider the example of plane gravitational waves emanating from a weak source. Since the gravitational field is weak, any self-interaction may be ignored and the metric tensor can be written as:

$$g_{\alpha\beta} = \eta_{\alpha\beta} + h_{\alpha\beta} \qquad (2)$$

where $\eta_{\alpha\beta}$ is the metric tensor for Minkowski (i.e., uncurved or 'flat') spacetime. Plane gravitational waves are represented by the tensor equation:

$$h_{\alpha\beta} = b_{\alpha\beta} \exp(ik_\mu x^\mu) + b^*_{\alpha\beta} \exp(-ik_\mu x^\mu) \qquad (3)$$

where $b_{\alpha\beta}$ is called the polarisation tensor, $b^*_{\alpha\beta}$ is its complex conjugate and k_μ is the wave vector. In the geometric view, these plane waves represent *ripples* in the 'fabric' of spacetime. If, however, equations (3) are quantized, the result is quantum fields of the form:

$$h_{\alpha\beta} = \sum_\sigma \int \{a(\mathbf{k},\sigma)\, b_{\alpha\beta}(\mathbf{k},\sigma) \exp(ik_\mu x^\mu) + a^\dagger(\mathbf{k},\sigma)\, b^*_{\alpha\beta}(\mathbf{k},\sigma) \exp(-ik_\mu x^\mu)\}\, d^3k$$

where $a(\mathbf{k},\sigma)$ and $a^\dagger(\mathbf{k},\sigma)$ are annihilation and creation operators respectively.[8] In the particle picture so obtained, the gravitational wave is considered to be made up of large numbers of quantum particles called gravitons of momentum $\hbar\mathbf{k}$ and helicity σ, where $\hbar$ is Planck's Constant divided by 2π. (The graviton is the particle that is responsible for mediating the gravitational force.) In this formalism, gravitons are created by application of the creation operator $a^\dagger(\mathbf{k},\sigma)$ to states described in a (mathematical) Hilbert space. It should be fairly obvious here that both particle and spacetime pictures cannot be (literally) correct. In the context of the realist assessment of theories of gravitation, the pivotal question surely is: *Which picture* (quantum particle or spacetime) *is warranted?*

GRAVITATIONAL TIME DILATION AND CLASSICAL GENERAL RELATIVITY

Gravitational time dilation is the phenomenon where time intervals are shorter when measured close to the source of a strong gravitational field than when measured further away from the source, i.e., at a 'higher' gravitational potential. (This is sometimes inaccurately referred to as the 'slowing down of clocks' in a gravitational field.) The gravitational redshift effect, where light travelling away from the source of a gravitational field suffers a change in wavelength towards the red end of the spectrum (and therefore a decrease in frequency) usually is associated with gravitational time dilation. Gravitational redshift has been the subject of empirical studies and is experimentally confirmed.[9]

The change in the frequency of light as it travels from one gravitational potential to another can be calculated using Newtonian gravitation and Einstein's photon hypothesis. Consider a photon of frequency f and kinetic energy $K = hf$, where h is Planck's Constant. The photon is emitted in a region with gravitational potential Φ and travels to another region with ('higher') gravitational potential Φ', e.g. from the Sun to the Earth. When this occurs the photon undergoes a redshift and increases its gravitational potential energy. As a consequence, it suffers a decrease in frequency and a loss of kinetic energy to $K' = hf'$, where f' is the frequency of the photon when observed at the 'higher' gravitational potential. The ratio of the two frequencies can be calculated by equating the total energies (kinetic plus potential) of the photon in the two regions: $(f'/f) = [(1 + \Phi/c^2) / (1 + \Phi'/c^2)] \approx (1 - \Delta\Phi/c^2)$, where the change in gravitational potential is $\Delta\Phi = \Phi' - \Phi$ (> 0) and c is the speed of light in vacuum. In deriving this result we have used the weak principle of equivalence and the assumption that both the emitter and the absorber of the light are of a non-relativistic nature.[10] The increase in the potential energy of the photon is due to the change in gravitational potential, but since its total energy remains constant the decrease in frequency is accountable *entirely* in terms of the loss of kinetic energy.

If we now compare the photon frequency f' with the frequency f^* of another photon emitted from an identical light source which is situated in the 'higher' gravitational potential (e.g. on the Earth), then we find (as expected) that $f' < f^*$ i.e., the light that has travelled from 'lower' to 'higher' gravitational potentials has been redshifted. These two frequencies are related by:

$$f^* = (1 + \Delta\Phi/c^2)\, f' \qquad (4)$$

This equation also may be derived using classical general relativity. Suppose that the components of the metric tensor are given by equation (2), i.e., $g_{\alpha\beta}$ differs only slightly from its Minkowski counterpart by the $h_{\alpha\beta}$. The tensor $h_{\alpha\beta}$ represents the deviation from 'flat' spacetime with $|h| << 1$. If we assume a static gravitational field and a frame of reference at rest (or moving with relative velocity $|v| << c$) with respect to the field, then the temporal component of the metric tensor $g_{\alpha\beta}$ is given by:

$$g_{oo} = (1 + 2\Phi/c^2) \qquad (5)$$

where Φ (> 0) is the (Newtonian) gravitational potential of the field and terms of order h^2 have been neglected as being too small to contribute physically.[11] The metric of spacetime is given by $ds = (g_{\alpha\beta}\, dx^\alpha\, dx^\beta)^{1/2}$, so that a clock stationary with respect to and outside the gravitational field will record a time:

$$d\tau = (g_{oo})^{1/2}\, (dx^o) \qquad (6)$$

where dx^0 is the coordinate time registered on a clock inside the field. Substituting equation (5) into equation (6) yields: $d\tau = (1 + 2\Phi/c^2)^{1/2} (dx^0)$. If $\Phi << c^2$, then $(1 + 2\Phi/c^2)^{1/2} \approx (1 + \Phi /c^2)$ and the time intervals recorded by these two clocks will be related by the equation: $\Delta\tau = (1 + \Phi/c^2)\ \Delta t$, where Δt is the time interval recorded inside the gravitational field. In the case of two clocks that both are situated in the field but at different distances from its source, the time intervals measured will be related by:

$$\Delta\tau = (1 + \Delta\Phi /c^2)\ \Delta t \tag{7}$$

where $\Delta\Phi$ is the change in gravitational potential and $\Delta\tau$ is the time interval recorded at the greater distance (i.e. in the 'higher' gravitational potential). Equation (7) is the equation of gravitational time dilation.

Since the time intervals measured in two different gravitational potentials are related by equation (7), the frequency of light measured from a light source further away from the field's origin will be greater than the frequency of light from an identical source which is close to the origin of the gravitational field. The frequencies of the light sources situated at different distances from the field's origin are related in the same manner as the time intervals $\Delta\tau$ and Δt, i.e., $f^* = (1 + \Delta\Phi/c^2)\ f'$, in agreement with equation (4). Indeed, the same result can be shown to follow from curved spacetime metrics without recourse to equation (2).[12] Thus, in classical general relativity, the gravitational redshift of light results solely from gravitational time dilation, i.e., the redshift is due only to differences in coordinate times found in curved spacetimes. With Newtonian mechanics and gravitation, the decrease in the frequency of light is attributable to a loss of kinetic energy. In classical general relativity, however, gravitational potential energy is not well defined and is not required to explain gravitational redshift.

Consider now the gravitational version of the asymmetric ageing of twins. It follows from equation (7) that if instead of clocks we have two identical biological specimens, one located in a 'higher' gravitational potential than the other, then the one in the 'higher' potential will physically 'age faster'. This could be demonstrated by initially having the specimens at the same spacetime location, separating them off to locations with different gravitational potentials and then bringing them back together. The discrepancy in the time intervals measured in the different gravitational potentials will be directly correlated with the lesser physical ageing of one of the pair of otherwise identical biological specimens. This phenomenon is unexplained in the context of a 'flat' spacetime as long as low velocities are used for the transport of the specimens. It is the curvature of spacetime that provides a natural explanation of the twin situation.

GRAVITATIONAL TIME DILATION AND THE PARTICLE APPROACH TO QUANTUM GRAVITY IN MINKOWSKI SPACETIME

Most particle theorists see gravity as another classical (force) field to be quantized just like the electromagnetic field has been. However, general relativity can be differentiated in several respects from field theories that have been quantized. In particular we may point to the following distinguishing features: the non-linearity of the field equations (1); the dynamic nature of the metric tensor; and the dimensionality of the coupling constant κ. These aspects pose formidable problems to quantization, problems not encountered when quantizing other classical fields. All attempts to produce a fully quantized theory of gravity have failed, but not for lack of effort. Despite the methods having been many and varied, none have proved successful notwithstanding the exhausting labours of theoretical physicists.[13]

In conventional quantum field theory, a given classical field is quantized on a 'flat' background (i.e., Minkowski spacetime). Likewise, the covariant quantization of the gravitational field can be attempted on a 'flat' spacetime background. This 'flat' background is obtained by assuming that the metric tensor differs from the Minkowski form by the factors $h_{\alpha\beta}$ only, i.e., the metric tensor is split as per equation (2). In this case though, the $h_{\alpha\beta}$ are interpreted as representing a self-interacting, massless, spin-2 field propagating in Minkowski spacetime, similar to the weak gravitational wave case. The quanta of this field are gravitons. The gravitational force, in this context, is explained by a continual exchange of 'virtual' gravitons in much the same way as electromagnetic forces are mediated by an exchange of 'virtual' photons. (The particles are 'virtual' because the limits of the Uncertainty Principle do not allow for their experimental detection.)

All theories of quantum gravity must yield the same macroscopic predictions as classical general relativity does, including the same quantitative result for gravitational redshift. In the covariant method, computations are made by specifying an action functional, as is routinely performed in quantum field theory. Where a photon moves from 'lower' to 'higher' gravitational potentials, this corresponds to an interaction between gravitational and electromagnetic fields. The action functional is found from the Einstein Lagrangian density $\mathcal{L}_G$ in conjunction with a Lagrangian density for the electromagnetic field $\mathcal{L}_{EM}$.[14] The total action then is defined by the integral $\int (\mathcal{L}_G + \mathcal{L}_{EM})\, d^4\boldsymbol{x}$. Graviton-photon interactions of low order are calculated by expanding the Lagrangian densities as power series and constructing the scattering matrix elements from the resulting action. (Such calculations are extremely tedious and will not be reproduced here.)[15] The *crucial* point, however, is that the explanation for the redshift of light in covariant quantum gravity is essentially the same as when Newtonian gravitation is applied to the photon hypothesis, i.e., the change in frequency of a photon is due to a loss of kinetic energy.

Whilst it is the case that the assumption of a coupling of gravitons to photons can be used to calculate the size of the gravitational redshift, the usual account of this redshift is that the decrease in photon frequency implies a difference in coordinate times. This inference is incorrect because the change in photon frequency results, not from time dilation, but from a loss of kinetic energy. Such energy loss does not explain the lesser physical ageing of one of a pair of otherwise identical biological specimens situated in different gravitational potentials. Therefore, the phenomenon of gravitational asymmetric ageing is not only unexplained, it has become *utterly mysterious*.

There are further rationales that can be made in favour of this conclusion:

(1) The loss of kinetic energy of the photon is a consequence of the law of conservation of total energy. Physical processes by which we measure time may be affected by the operation of this law, but time itself is not.

(2) Whilst it may be the case that some physical processes may be affected by a loss of kinetic energy from the system in question, this (in isolation) does not warrant the further inference that the actual ageing processes of biological organisms would be altered.

(3) If the physical ageing of objects were affected by a loss of kinetic energy then this would be apparent in other situations, which it is not. Consider a simple example — the removal of heat from an object (as its heat energy ultimately is the sum of the individual molecular kinetic energies). Suppose that the object is a pure chemical system, a vat of liquid reactants say. A reduction of the heat content of the vat would produce slower and slower chemical reactions. If time intervals were to be judged by the 'rate of reaction' in the vat, then an inference about time similar to that made about frequency decrease in the 'flat' spacetime case might be made. Clearly though, any conclusion about time based on rates of reaction would be nonsense, for we know from the study of chemical kinetics that reaction rates vary enormously and depend on all sorts of different factors (e.g. temperature, pressure, chemical concentrations, surface areas, catalysts, etc.).

(4) If the particle theorists were correct, then the loss of a large amount of kinetic energy from one of a pair of otherwise identical biological organisms would produce a difference in their respective physical ages. Since heat energy is just the sum of molecular kinetic energies, this loss of energy could be achieved by progressively cooling one of the pair. Suppose this were to be done, what would happen? The biochemical reactions of the organism that is being cooled certainly will slow down. This, however, is a situation no different in principle from that depicted in point (3) above.

No sound inference about time itself follows from any of these considerations.

The twin situation demonstrates that gravitational time dilation is not explained by graviton interactions on a 'flat' spacetime background. This conclusion has (at least) two important consequences:

(i) spacetime curvature cannot be dispensed with merely by postulating gravitational exchange quanta; and

(ii) theories of quantum gravity need to incorporate a (geometrically interpreted) curved spacetime.

Theoreticians have pursued conventional quantization methods on 'flat' spacetime backgrounds since the early 1930s. However, even at an intuitive level, it should have been realised that these methods are unsuitable because they cannot account properly for gravitational time dilation. These 'flat' spacetime methods finally are being given away for the reason that they produce non-renormalizable theories (i.e., theories with infinities that cannot be removed).[16] Yet, the failure to account properly for gravitational time dilation always has been a sufficient reason to dismiss any relativistic theory of gravity with an uncurved spacetime.

THE PARTICLE APPROACH TO QUANTUM GRAVITY IN CURVED SPACETIME

In theories of quantum gravity, the continual exchange of 'virtual' particles accounts for the force of gravity. In order to account for gravitational asymmetric ageing, theories of quantum gravity need to incorporate a (geometrically interpreted) curved spacetime, as noted above. Does it really make sense to postulate gravitational exchange quanta in a theory that already is endowed with a curved spacetime? Such theories have a somewhat confusing, and perhaps incompatible, mix of basic entities (i.e., gravitons in a curved spacetime). What is important to note here is that, in addition to accounting for gravitational time dilation, curved spacetime readily explains deviations from inertial motion due to the influence of massive bodies. In other words, spacetime curvature can account, of itself, for both the 'force' of gravity and asymmetric ageing without the need for exchange quanta. (Gravitational force, as such, does not exist in a curved spacetime — it is fictitious, like centrifugal force.) Therefore, gravitational exchange quanta are *superfluous* in a curved spacetime for these purposes.

Another result of taking a particle view of gravity in curved spacetime has been shown by von Borzeszkowski and Treder who use a standard covariant technique. We shall reproduce part of their argument. Let the metric tensor be decomposed as follows:

$$g_{\mu\nu} = \delta_{\mu\nu} + e h_{\mu\nu} \tag{8}$$

where $\delta_{\mu\nu}$ is a fixed, but not necessarily 'flat' background metric and $e << 1$. (Negligible back reaction of the $eh_{\mu\nu}$ fields on the curvature of $\delta_{\mu\nu}$ is assumed.) Let L and Ω be characteristic lengths over which the δ and h parts of the field might (conceivably) alter in a substantial manner. The Ricci tensor $R_{\alpha\beta}$ formed from this metric (equation (8)) then can be expanded in powers of e:

$$R_{\alpha\beta}(\delta_{\mu\nu} + eh_{\mu\nu}) = R_{\alpha\beta}(\delta_{\mu\nu}) + eR'_{\alpha\beta}(h_{\mu\nu}) + e^2 R''_{\alpha\beta}(h_{\mu\nu}) + e^3 R'''_{\alpha\beta}(h_{\mu\nu}) + \ldots \quad (9)$$

where $R_{\alpha\beta}(\delta_{\mu\nu})$ is the Ricci tensor for the background metric.[17] The terms R', R'' and R''' have orders of magnitude L^{-2}, $e^2 \Omega^{-2}$, $e^3 \Omega^{-2}$ respectively. The powers of e will estimate each of the these terms only when $L \geq \Omega$. It follows that only for low frequency fields $h_{\mu\nu}$ will the background metric be governed by the equation $R_{\alpha\beta}(\delta_{\mu\nu}) = 0$. If $R_{\alpha\beta}(\delta_{\mu\nu})$ is made equal to zero for all Ω, then equations (9) will not yield the field equations (1). If we are to remain within the framework of general relativity, as distinct from another theory of gravity, then this approach can be applicable only to low frequency (i.e., classical) fields.[18]

Since the quanta of the $h_{\mu\nu}$ fields are gravitons, it is inferred from the above considerations that the quantum representation is an approximate one which is valid only when low frequency gravitational phenomena are involved. The conclusion reached on this basis is that the production of field quanta by covariant quantization does not correspond to any real, physical process.[19] Whilst it is obvious that the covariant method does not provide a full theory of quantum gravity, this result does raise the important question of whether the process of quantization in particle approaches will correspond to anything real (and measurable). Also, the problem of divergent terms in perturbation expansions does not go away when the background spacetime is curved.[20]

It has been the belief that classical gravity is a field *on par* with other classical fields that has lead particle theorists to assert that the gravitational field not only ought to be, but indeed must be quantizable as other fields have been. This view is mistaken, as classical force fields play out their interactions within a spacetime structure. The gravitational field should be considered a classical field for mathematical purposes only. Treating it otherwise has been an open invitation to all sorts of technical and conceptual problems. Part of the explanation for the existence of these problems is what was earlier referred to as the application of theories/laws developed in one sphere of applicability being applied to a different sphere without due consideration of the appropriateness of such application and its consequences.

In sum, given that the graviton representation is only an approximate one which does not lead to any new measurable effects, is superfluous in a curved spacetime and remains perturbatively non-renormalizable, then there would seem to be little advantage in pursuing particle approaches to quantum gravity.

OTHER THEORIES OF QUANTUM GRAVITY

There's not much joy to be found in other quantum gravity theories either. We shall look briefly at the major problems found in two main theoretical contenders. Canonical theories of quantum gravity are those based on a Hamiltonian formulation of the classical theory. In these canonical approaches, the metric tensor itself is quantized! But if such a procedure is performed, how is the quantized metric tensor to be interpreted? What will be its resulting ontological status? Does it remain a geometrical entity? A quantum 'particle' of some sort? Some bizarre hybrid of geometrical and particle natures? Or something altogether different again? The conceptual 'pictures' that emerge from canonical quantum gravity theories are *not unambiguous* and require more precise renderings than have been advanced by particle theorists.

One of the most serious difficulties in canonical theories of quantum gravity is the problem of time (as it is called). In these theories, time is 'eliminated' from the governing equation of evolution of the relevant system, i.e., the Schrödinger equation is modified to one not involving time. However, in the low energy limit, it is required that state functions do evolve in time. Thus some internal degree of freedom must be used in place of parameter time against which state functions can 'evolve'.[21] Moreover, such an internal variable would have to result in the proper time parameter in the low energy limit. It is not always clear what this internal variable should be or how it is to be chosen. Ashtekar summed up the problem of time when he wrote:

> If there is no classical time, however, how can one do physics? After all, the business of physics is to make predictions. How can one predict if one does not have access to the familiar notion of time? [22]

String theory is the extension of covariant methods to ultra-small one-dimensional objects called (not surprisingly) strings. In some versions of string theory spacetime is not taken as a basic feature at all. The different modes of vibration of the string are identified as quantum particles, but the string itself does *not exist in* anything akin to a classical spacetime. Instead, classical spacetime is (somehow) supposed to emerge from a more fundamental realm as a low energy approximation.[23] What this fundamental realm is supposed to be has not been defined. The loss of a familiar notion of spacetime as both an arena of and participant in events has the potential to lead to serious problems, especially those associated with time. For example, such theories face a major problem in providing an *explanatory* account of gravitational asymmetric ageing, as distinct from being able to predict the correct numerical value for gravitational redshift.

It is generally agreed that there is something different about spacetime structure at and below scales of the order of the Planck Length ($\approx 10^{-33}$ cm.). This, however, does not require the removal of spacetime. Rather it would tend to suggest that the classical concept of an infinitely differentiable manifold with its defined metric as a model of spacetime is inappropriate at this scale and that an some other *geometrical* structure is required. We shall return to this point in the next section.

String theory suffers from a complaint that might be called 'ontological proliferation'. First, most versions require the number of spatial dimensions to be greater than three. This is dictated by the requirements of conformal invariance, i.e., the invariance of the null-cone structure. These unseen dimensions (which may be as many as twenty-six!) are concealed by being 'compacted' into circles with radii of the order of the Planck Length. Second, the types and numbers of hypothetical particles also are increased markedly in string theory. The reasons for embarking on such complexities are purely theoretical and are justified by appeal to certain desiderata: the cancellation of infinities; the unification of fundamental particles and interactions; and the production of aesthetically satisfying theories.[24] However, theories displaying such 'ontological overkill' cannot be entertained seriously for an *indefinite* period without very convincing experimental evidence. There would appear to be very little likelihood of ever gaining this evidence due the enormous energies required to test such theories.[25] Given these factors, one is inclined to side with the physics Nöbel laureate, Sheldon Glashow, who has stated that he is "waiting for the superstring to break" !

CAN GRAVITY BE FULLY QUANTIZED?

There are severe problems of both a conceptual and a technical nature encountered in trying to construct a complete, physically viable, self-consistent, empirically adequate theory of quantum gravity. This point is not disputed. Purely in view of the scope and magnitude of these problems one might be tempted to conclude that the gravitational field cannot be quantized rigorously. Perhaps general relativity and quantum mechanics really are (to use Isham's phrase) "intrinsically incompatible".[26] After all, there is no *a priori* reason for asserting that gravity must be a quantum field. An alternative view is that the failure to solve the obstacles of quantum gravity is due to a lack of ingenuity on the part of theoreticians. Regardless of one's opinion, the reasons for seeking a theory of quantum gravity (points (I)–(IV) in the introduction) still need to be properly answered (or perhaps 'disarmed' is a more appropriate description).

Consider now a tentative (if not somewhat speculative) response. Equation (1) does demand a better form of 'accord' between gravitational and matter fields. The simplest unity is a form of semi-classical quantum gravity. The Einstein tensor can be equated with the expectation value of the quantum operator formed from the stress-energy-momentum tensor, viz.:

$$G_{\alpha\beta} = - \kappa <T_{\alpha\beta}> \quad (10)$$

Difficulties with this semi-classical approach have been indicated in a number of articles.[27] The alleged problems include violation of the Uncertainty Principle, the possibility of faster-than-light communication and violation of momentum conservation. Some solutions to problems connected with semi-classical quantum gravity have been dealt with. In particular, if a version of quantum mechanics in which the wavefunction does not collapse is accepted then these problems do not occur.[28] This suggests that at least part of these difficulties might lie with 'orthodox' quantum theory. Non-collapse versions would not only deal with the above problems, but would go a long way in avoiding most of the standard quantum paradoxes.

In any case, equations (10) are still formally inconsistent. Dirac pointed out that it is mathematically incorrect merely to equate the Einstein tensor with the expectation value of the stress-energy-momentum operator. $G_{\alpha\beta}$ is a Hilbert space scalar, whereas $\mathbf{T}_{\alpha\beta}$ depends on the choice of the state vector in a Hilbert space. In order to avoid inconsistencies, such as those mentioned above, the latter quantity must be equated with the *average* value of the Einstein tensor:[29]

$$<G_{\alpha\beta}> = - \kappa <T_{\alpha\beta}> \quad (11)$$

There is no actual quantization of the gravitational field needed here and the classical geometric flavour is preserved. This deals with quantization reason (I) in the introduction. Also, given the problem of measurement in quantum gravity, it is highly probable that the above semi-classical treatment will be epistemically equivalent to any other. This is not to suggest, however, that a version of semi-classical quantum gravity is the complete picture or that all the problems of quantum gravity are solvable within such a framework. Semi-classical quantum gravity, for example, does not resolve the question of the existence of spacetime singularities.

A fundamentally geometric nature for gravitation would mean that a completely rigorous unification of all fields is *not* possible. Evidence in favour of the geometric nature of gravity and for the validity of the strong principle of equivalence continues to mount.[30] Acceptance of the geometric nature disposes of quantization reason (III). A geometric basis is not only consistent with the arguments presented in this essay, but also would account for the *insurmountable* difficulties encountered by particle theorists in attempting to quantize the classical gravitational field equations in artificial ways. Given that all attempts to create a full quantum theory of gravity have failed, it might prove more productive to concentrate on strategies that preserve the geometric character of gravity rather than on mathematical methods of 'quantizing' the field equations.

Consider what it would signify if spacetime naturally existed in a form which has a smallest unit, i.e., a minimum four-dimensional hypervolume. Then, instead of spacetime being modelled as a manifold with a metric ranging over the real numbers, we would have a discrete spacetime with a 'metric' that ranges over the integers only. The size of these spacetime units would be so minute that even at moderately small distances, spacetime would have the appearance of a manifold and the semi-classical equations (11) would make an excellent approximation. The existence of a fundamental spacetime hypervolume also implies that there should be some minimum spacetime interval that would be an invariant quantity from one frame of reference to another.

Moreover, if this notion of discrete spacetime is correct then the traditional mathematical tool of differential geometry would have to be replaced. It is interesting to note that some theoreticians are starting to come around to such a view, as the following statement by C.J. Isham indicates:

> Perhaps the entire paraphernalia of differential geometry is only appropriate at scales greater than the Planck length. But if so, with what should it be replaced ...? [31]

One possible answer to Isham's question is as follows: the spacetime continuum could be replaced with a metrically discrete one and the field equations (1) with finite difference equations. Indeed, the famous theoretican T.D. Lee, has made similar suggestions. He argues that difference equations are more fundamental than differential ones and that the latter should be seen only as approximations.[32]

The idea that spacetime might be discrete is not new. Various attempts have been made to construct viable equations that represent phenomena in a (metrically) non-continuous spacetime, usually with the stated aim of removing the infinities that plague field theories.[33] There have been a number of problems found, but the most serious would appear to be the violation of Lorentz invariance.[34] Whilst it may be the case that continuous Lorentz transformations only hold approximately if spacetime is discrete, their violation is cause for concern and if possible, should be avoided. A discrete spacetime (of the sort envisaged) would eliminate the problem of measurements to arbitrary accurary (quantization reason (II)). If a minimum four-dimensional hypervolume exists it would not be physically possible to make measurements to an accuracy less than the size of this hypervolume, for that would imply further divisibility. It should be noted, however, that a discrete approach to gravitation would bring its own assortment of difficulties, but ones no more problematic than those experienced thus far in the search for a correct theory of quantum gravity.

In conclusion, we note that geometrically interpreted curved spacetimes cannot be done away with by postulating gravitational exchange quanta and that particle based interpretations of quantum gravity bring forth a host of technical, conceptual and philosophical problems, which indicate their unsuitability. It is fair to say that there still remains much to investigate in the search for a complete theory of quantum gravity and discrete geometric models offer a promising alternative to the standard quantization ones.

University of Adelaide, South Australia

NOTES

* I would like to thank Graham Nerlich, John Forge and Peter Szekeres for helpful comments.

1 The reader is referred to the essays by Forge and Heathcote in this volume.
2 Cf. Rindler (1977), p.21; Wald (1984), p.68; Doughty (1990), pp.482-4.
3 McMullin (1984), p.14.
4 Cushing (1982), pp.43 & 78.
5 Isham (1988), p.82.
6 Nerlich and Westwell-Roper (1985), p.128.
7 Cf. Penrose (1982) and Isham (1984).
8 Weinberg (1972), p.288.
9 For example, see Will (1986), pp.54-64.
10 Weinberg (1972), pp.84-85.
11 Angel (1980), p.200.
12 See Wald (1984), p.137; or Weinberg (1972), pp.79-85.
13 The reader interested in these issues is referred to: Duff (1975); Duff and Toms (1984); Isham (1981, 1984, 1991); Kuchar (1981); Taylor (1989); Weingard (1989); Ashtekar and Stachel (1991).
14 $\mathcal{L}_G = (1/k)\ (-g)^{1/2}\ R$ and $\mathcal{L}_{EM} = -\ \frac{1}{4}\ F_{\alpha\beta}\ F^{\alpha\beta}$, where $F^{\alpha\beta}$ is the electromagnetic energy-momentum tensor, $g = |g_{\alpha\beta}|$, R is the Ricci scalar and k is a constant.
15 For technical details, see Duff (1975), pp.87-90.
16 Ashtekar (1991a), pp.7-8.
17 $R'_{\alpha\beta}\ (h_{\mu\nu}) = \delta^{\theta\tau}\ (h_{\theta\tau|\alpha\beta} + h_{\alpha\beta|\theta\tau} - h_{\tau\alpha|\beta\theta} - h_{\tau\beta|\alpha\theta})$ with (|) being covariant differentiation with respect to the background metric. Similar expressions apply for R'' and R''' terms, etc.
18 von Borzeszkowski and Treder (1988), pp.58-61.
19 The full non-linearity also can be taken into account. See von Borzeszkowski and Treder (1988), pp.62-66, for details.
20 Ashtekar (1991a), p.5.
21 Ashtekar (1991b), p.413.
22 Ashtekar (1991c), p.195.
23 For example, see Horowitz (1991), p.299.
24 Green (1985).
25 't Hoof (1979), p.328.
26 Isham (1981), p.2.
27 For example, Eppley and Hannah (1977).
28 Page and Geilker (1982), p.13.
29 von Borzeszkowski and Treder (1988), pp.1-2.
30 See Will (1981), chap. 8.
31 Isham (1991), p.359.
32 Lee (1986), p.198.
33 For example, Snyder (1947); Flint (1948); Hellund and Tanaka (1954).

[34] Cf. Schild (1948); Hill (1955); and Pavlopoulos (1967).

REFERENCES

Angel, R.B. (1980) *Relativity: The Theory and its Philosophy*. Pergamon, Oxford.
Ashtekar, A. (1991a) 'Introduction: The Winding Road to Quantum Gravity' in Ashtekar, A. and Stachel, J. (eds.) (1991).
— (1991b) 'Old Problems in the Light of New Variables' in Ashtekar, A. and J. Stachel (eds.) (1991).
— (1991c) *Lectures on Non-Perturbative Canonical Gravity*. World Scientific, Singapore.
Ashtekar, A. and Stachel, J. (eds.) (1991) *Conceptual Problems of Quantum Gravity*. Birkhauser, Boston.
Cushing, J. (1982) 'Models and Methodologies in Current Theoretical High-Energy Physics', *Synthese* 50, 5-101.
Doughty, N.A. (1990) *Lagrangian Interaction: An Introduction to Relativistic Symmetry in Electrodynamics and Gravitation*. Addison-Wesley, Sydney & New York.
Duff, M.J. (1975) 'Covariant Quantization' in Isham, C.J., Penrose, R. and Sciama, D.W. (eds.), *Quantum Gravity: an Oxford Symposium*. Clarendon, Oxford.
Duff, M.J. and Toms, D.J. (1984) 'Divergences and Anomalies in Kazula-Klein Theories' in Markov, M.A. and West, P.C. (eds.), *Quantum Gravity*. Plenum Press, New York and London.
Eppley, K. and Hannah, E. (1977) 'The Necessity of Quantizing the Gravitational Field', *Foundations of Physics* 7, 51-68.
Flint, H.T. (1948) 'The Quantization Space and Time', *Physical Review* 74, 209-10.
Green, M.B. (1985) 'Unification of Forces and Particles in Super-string Theories', *Nature* 314, 408-14.
Hellund, E.J. and Tanaka, K. (1954) 'Quantized Space-Time', *Physical Review* 94, 192-95.
Hill, E.L. (1955) 'Relativistic Theory of Discrete Momentum and Discrete Space-Time', *Physical Review* 100, 1780-83.
Horowitz, G.T. (1991) 'String Theory Without Space-Time' in Ashtekar, A. and Stachel, J. (eds.) (1991).
Isham, C.J. (1981) 'Introduction to Quantum Gravity' in Isham, C.J., Penrose, R. and Sciama, D.W. (eds.), *Quantum Gravity 2: A Second Oxford Symposium*. Clarendon, Oxford.
— (1984) 'Quantum Geometry' in Christensen, S.M. (ed.), *Quantum Theory of Gravity: Essays in honor of the 60th birthday of Bryce DeWitt*. Adam Hilger, Bristol.
— (1991) 'Canonical Groups and the Quantization of Geometry and Topology' in Ashtekar, A. and Stachel, J. (eds.) (1991).
Kuchar, K. (1981) 'Canonical Methods of Quantization' in Isham, C.J., Penrose, R. and Sciama, D.W. (eds.), *Quantum Gravity 2: A Second Oxford Symposium*. Clarendon, Oxford.
Lee, T.D. (1986) 'Physics in Terms of Difference Equations' in de Boer, J., Dal, E. and Ulfbeck, O. (eds.), *The Lesson of Quantum Theory*. Elsevier, Amsterdam.
McMullin, E. (1984) 'A Case for Scientific Realism' in Leplin, J. (ed.), *Scientific Realism*. University of California Press, Berkeley and Los Angeles.
Nerlich, G. and Westwell-Roper, A. (1985) 'What Ontology Can Be About: A Spacetime Example', *Australasian Journal of Philosophy* 63, 127-42. Reprinted in Nerlich's *What Spacetime Explains: Metaphysical Essays on Space and Time*. Cambridge University Press, Cambridge, 1994.
Page, D.N. and Geilker, C.D. (1982) 'An Experimental Test of Quantum Gravity' in Duff, M.J. and Isham, C.J. (eds.), *Quantum Structure of Space and Time*. Cambridge University Press, Cambridge.
Pavlopoulos, T.G. (1967) 'Breakdown of Lorentz Invariance', *Physical Review* 159, 1106-10.

Penrose, R. (1982) 'Some Remarks on Gravity and Quantum Mechanics' in Duff, M.J. and Isham, C.J. (eds.), *Quantum Structure of Space and Time*. Cambridge University Press, Cambridge.

Rindler, W. (1977) *Essential Relativity: Special, General, and Cosmological*. Springer, Berlin.

Schild, A. (1948) 'Discrete Space-Time and Integral Lorentz Transformations', *Physical Review* 73, 414-15.

Snyder, H.S. (1947) 'Quantized Space-Time', *Physical Review* 71, 38-41.

Taylor, C.C. (1989) 'String Theory, Quantum Gravity and Locality' in Leplin, J. and Fine, A. (eds.), PSA 1988: *Proceedings of the 1988 Biennial Conference of the Philosophy of Science Association* (Vol. 2). Philosophy of Science Association, East Lansing, Michigan.

't Hooft, G. (1979) 'Quantum Gravity: A Fundamental Problem and Some Radical Ideas' in Lévy, M. and Deser, S., *Recent Developments in Gravitation*: Cargèse 1978. Plenum, New York.

von Borzeszkowski, H.-H. and Treder, H.-J. (1988) *The Meaning of Quantum Gravity*. Reidel, Dordrecht.

Wald, R.M. (1984) *General Relativity*. Chicago University Press, Chicago.

Weinberg, S. (1972): *Gravitation and Cosmology: Principles of the General Theory of Relativity*. Wiley, New York.

Weingard, R. (1989) 'A Philosopher Looks at String Theory' in Leplin, J. and Fine, A. (eds.), PSA 1988: *Proceedings of the 1988 Biennial Conference of the Philosophy of Science Association* (Vol. 2). Philosophy of Science Association, East Lansing, Michigan.

Will, C.M. (1981) *Theory and Experiment in Gravitational Physics*. Cambridge University Press, Cambridge.

— (1986) *Was Einstein Right? Putting General Relativity to the Test*. Basic Books, New York

D.M. ARMSTRONG

COMMENTS ON LIERSE

Caroline Lierse criticizes David Lewis and myself for seeking to uphold the Humean Distinct Existences principle — no necessary connection between distinct existences — but at the same time covertly abandoning it in our discussion of properties. She devotes most of her space to Lewis, who can speak for himself. I plead guilty to using some infelicitous language in my 1978 book on universals, to which she refers, but think I am innocent of the charge she brings.

In that book I advocate, and would still advocate, causal identity-conditions for universals. Properties bestow powers on particulars, the same property always bestows the same powers, and different properties bestow different powers. My detailed discussion is in Vol.2, Ch.2, Sec.1, which Lierse refers to only as an afterthought in her fn.10. What I should have said, and indeed what I believed then and still believe, is that I took these suggested identity-conditions to be contingent. This would entail the *possibility* of powerless (epiphenomenal) properties, and of different properties having the same causal powers. This view naturally leads on to the view that laws of nature are contingent, although I reject Humean theories of laws.

What I did say was that the suggested identity-conditions could only be "pragmatically justified", and that we could *conceive* of properties without these identity-conditions. I did offer some explication of what "pragmatically justified" meant in the text immediately following. I said for instance that it is hard to see how we could ever become aware of an epiphenomenal property (pp.44-5). This, I think, ought to have suggested to Lierse that my identity-conditions were not intended to be necessary truths.

I made points like these to Lierse in correspondence with her after I had read an earlier draft of her paper. It seems, though, that she does not believe me! And it is true that I am an interested party and may have skewed my memory of what I used to think then. But she is an interested party also.

I am aware, of course, that my view that properties are only contingently connected with the causal powers they bestow faces difficulties, quite serious difficulties. Lierse gestures towards an important argument when she says that my sort of view "... is left with an ontology consisting of properties whose natures are *ex hypothesi* inscrutable" (p. 31). But as Marshal Foch was wont to observe, "It rains on the enemy too"! In a book now completed but still be published, *A World of States of Affairs*, I discuss two conceptions of properties, calling them the Categoricalist (my own view, along with many other philosophers) and the Dispositionalist theories (Lierse, along with many other philosophers). I find

P.J. Riggs (ed.), Natural Kinds, Laws of Nature and Scientific Methodology, 227–228.

difficulties, serious difficulties, for both views, and difficulties also for an intermediate position. My (completely disputable) conclusion is that, on balance, the Categoricalist view is to be preferred.

ADRIAN HEATHCOTE

COMMENTS ON FORGE

In his essay 'Laws and States in Quantum Mechanics', John Forge presents a case for considering laws of nature to be privileged sets of states, trajectories in the quantum mechanical analogue of phase space. Having presented an argument to show that states have to be taken with full ontological seriousness, Forge then uses those states to undergird his favourite account of laws and explanation – called the *Instance View*. On this view laws are a special sort of pattern, a certain kind of regularity, and explanations are instances of these regularities (so that you explain some phenomenon A by indicating that it is an instance of some regularity R). Forge then uses this account of laws and explanation to urge the acceptance of van Fraassen's Modal Interpretation of Quantum Mechanics.

This alliance between van Fraassen and Forge is, on one level, not at all surprising, for the instance view of explanation is partly motivated by a mistrust of *causal explanation* — since the latter does not readily mesh with the supposed absence of causal connectedness in quantum mechanics (an absence that van Fraassen has defended). Seen this way, Forge and van Fraassen share a neo-Humeanism mistrust of natural necessity, preferring instead a *sub species aeternitatas* view of physical systems evolving in phase space: mere regularities, but continuous regularities for all that. But even though Forge and van Fraassen share a certain sympathy with classical Humeanism, the former is a Realist of some stripe while the latter is a quite thorough-going anti-Realist, and this, I shall argue, is all the difference that is required to make quantum mechanics a considerable problem for Forge's views of laws and explanation.

In fact Forge's conscription of quantum mechanics (QM) faces two problems, one that he tries to meet and one that he does not. The problem that he tries to meet is the measurement problem and Forge's proposal is that we embrace the Modal Interpretation. Forge rightly feels that he must say something about the measurement problem since it is this that stands squarely in the way of a consistent interpretation of the theory and only a consistent interpretation can be taken with full ontological seriousness. In the Modal Interpretation of QM the dynamical state of the system undergoes no von Neumann collapse on measurement and therefore evolves entirely in accordance with the Schrödinger equation. It might seem that this would make quantum mechanics non-probabilistic, because the Schrödinger equation is deterministic. Not so, say van Fraassen and Forge, the probabilistic character of QM is ineliminable but it is not a probabilistic change of the dynamical state; instead, the probabilities emerge from the value shown by the measuring instrument as a result of its interaction with the measured system. The mixture of apparatus states and system states is a bundle of possibilities

P.J. Riggs (ed.), Natural Kinds, Laws of Nature and Scientific Methodology, 229–233.

until one drops into actuality with a certain probability. On van Fraassen's view our quantum reality is a world of mere possibilities, *potentia*, until one becomes actual at a particular time and place.

This may make van Fraassen's metaphysics sound distinctly bizarre, but in truth his view is considerably stranger than even this summary suggests. The 'oddness' of van Fraassen's view arises from the fact that the state of the system — strictly, the dynamical state — undergoes no probabilistic collapse but that something else — the value state — does. But if this is so then the dynamical state and the value state must be continually coming apart. This is, in fact, the case. On van Fraassen's view a composite system in a pure state evolves deterministically and the value state is completely determined by the dynamical state. The parts of the composite system may, however, be in mixed states and in this case the value states of the parts need not be determined. This means that the whole system can consist of a definite value state *for the whole* while its constituents are left indeterminate — and when we remember just how ubiquitous mixed states are (since all systems are continually interacting with background fields) then the only entity that genuinely has fixed states is the universe in its entirety: everything less than this will be an indeterminate part of a larger indeterminate part.

The dynamical state therefore evolves in a continuous fashion, but what one is capable of observing, the value state, simply seems to drop out of some aleatory heaven. Having dropped out thus, it is completely epiphenomenal, for the dynamical state simply goes on evolving deterministically regardless of how the system has temporarily appeared. As I have put it elsewhere, on van Fraassen's view quantum systems are like incompetent motorists, continually signally their intention to turn left or right but then driving straight on ahead. But it is even worse than that: passers-by see the driver turn left or right even though no turn is ever made – and a moment later they will see that the driver is signalling the other way! On the spectrum of Realist positions van Fraassen's view is at the extreme *Unrealist* end.

On the Modal Interpretation the value state of a component system in a mixed state is a possibility that becomes actual at a particular moment, but drops back into mere possibility a moment later. Since the value state cannot affect the dynamical state of the system, a subsequent observation will quite likely yield some other value (which is to say that the emergence of a value does not signify anything other than a momentary collapse in the value state of the system). There is thus no guarantee that our observations will be consistent with one another from one moment to the next, and everything to say that they will not. Van Fraassen tries to meet this problem by claiming that observations made on observers (second-order observations) will find their observations consistent, even though it is highly probable that their observations will be inconsistent (at the first-order level)! But why? The dynamical state of the observer's brain is itself simply evolving deterministically — how is it capable of having its own value states register on itself in a random fashion, when it itself is fully determined in its evolution? Either the value states cannot register on the brain, and it evolves regardless, or they do register and its deterministic evolution is interrupted. Either way, van Fraasasen's theory seems to run into a logical contradiction.*

Van Fraassen's metaphysic is thoroughly non-realist, and for a very good reason: if the Modal Interpretation is not to appear completely implausible van Fraassen must drastically limit the kind of questions that can be put to it. Indeed we are meant to limit ourselves just to asking whether it gives us the correct results for our observations — or sometimes just observations made on observations (and as we've just seen, even this restriction may not be enough to save it). So how, one wonders, does the realist Forge thinks that he can buy into the view? Surely the more real we make the dynamical and value states the more we will be getting back to the sudden probabilistic collapse (albeit, momentary) that van Fraassen's view is supposed to avoid. With this sudden probabilistic collapse will come all of the other demons of the measurement problem. Van Fraassen thinks that he can avoid these problems by relegating the awkward bits of the ontology to the non-actual, but what is left of this when it is *all* actual?

Of course it might be suggested that there are indicators of genuine ontological commitment in van Fraassen's presentation of his view and that these should be carefully considered alongside his more explicitly professed anti-realism. It is of course true that there are such signs — it is, after all, very difficult for an author to tell us how things are without seeming to tell us how things *really* are. But though it is true that van Fraassen has trouble putting his anti-realist metaphysics across, that alone could not make a realist interpretation of his view assured. It is not for nothing that van Fraassen warns us that a realist interpretation of the Modal view has quite bizarre consequences (and the non-realist interpretation is surely bizarre enough to be going on with!). Forge must somehow find a way of keeping van Fraassen's thoroughly non-realist explanatory structures, including his transitions from the possible to the actual, while at the same time changing the ontology so that the transitions are a genuine change of state of some kind. All of this looks like a very tall order, to say the least.

Forge is driven into this difficult impasse by a noble motive, the desire to solve the measurement problem. He wishes there to be only a single dynamical law so that quantum mechanics can be made to fit the structural picture that he finds in classical mechanics: the laws of the theory picking out privileged trajectories in the quantum mechanical analogue of phase space. Thus in section three we find that the standard account of quantum mechanics fails to deliver the right trajectories (the right ones being designated L(Q)); we then have a list of four possible ways that we can go (roughly four ways of interpreting QM) to try to bring the theory into line.

The four alternatives that Forge presents are, however, a rather mixed bunch and the statements of the views, as well as the subsequent arguments against them, are rather sketchy and unclear. What does emerge very clearly, though, is that the last of these is the preferred view:

(4) Deny that the dynamic state ever collapses and hence affirm that L(Q) holds in all circumstances.

This is to be preferred, it should be noted, over the first option:

(1) Deny that measurement interactions are lawful.

Up to this point, then, we know what Forge is for, even if we might not be certain that we know exactly what he is against: he is opting for a 'no-collapse' interpretation of the measurement problem and one with a single dynamics in which 'L(Q) holds in all circumstances'. We know, moreover, that Forge believes that the Modal Interpretation fits the bill.

However, by the time we reach the end of the paper, and after Forge has given a nicely encapsulated account of the Modal Interpretation, it is no longer quite so clear that this interpretation really does fit with his preferred option (4). It is certainly true that the state, called by van Fraassen the dynamical state, does not collapse, but the value state has its own dynamics and, even though it is certainly non-continuous, it is a dynamics that someone who is a Realist about the theory must take seriously. Forge confronts this problem head-on and promptly recognizes the need for a probabilistic law governing the propensity of system in a mixed state to have some value. Forge *must* take this course because to fail to do so would leave the results of measurements non-lawful — exactly the situation that was described and rejected in his option (1).

Now Forge is left with a seeming inconsistency in his account, for, recognizing the need for a probabilistic law now means that L(Q) is not the sole dynamical law for quantum systems, there is also the new L(MV) which governs its own proprietary state, the so-called value state — and whereas L(Q) does not collapse L(MV) does. (As I pointed out above, the value state undergoes *momentary* collapse whenever measurements are made; it would be mere semantic quibbling to think that this transition from the possible to the actual is not a kind of collapse.) There seems very little left of the neat picture that motivated the whole discussion in section three and that led Forge to embrace the Modal Interpretation in the first instance. It is as though, in giving his account of the Modal Interpretation, Forge has reluctantly become aware that the value state and its transitions will have to be taken seriously, that a Realist cannot afford to neglect them in favour of the, rather deceptively named, dynamical state. I think, however, that Forge should have abandoned his project at this point, noble though it was.

Not only is the Modal Interpretation not 'tuned' for the kind of realist projects that Forge has in mind, but trying to make it so will introduce just the kind of ontological and explanatory structures that send it out of line elsewhere — thus making it unsuitable all over again. Not only are there problems trying to bend van Fraassen's empiricist account into the more intricate shapes required for Realism, so there is also the unmet problem of the self-consistency of the Modal Interpretation. Forge is not obliged to solve that problem, or even address it, but with a view as bizarre as the Modal Interpretation it is probably worth pausing over – particularly if one is a principled realist, as Forge is.

The University of Sydney, New South Wales

* There are more objections made of the Modal Interpretation in my Critical Notice of van Fraassen's book in *Metascience*, New Series, Issue Three, 1993, pp.65-76.

KEITH HUTCHISON

COMMENTS ON THOMASON

One of the clear targets of Thomason's paper is the Feyerabendian portrait of Galileo as epistemic opportunist, hastening to substitute rhetoric for reason. Thomason reveals that Feyerabend has fallen into that awkward trap all critics must fear: when we claim to detect blemishes of logic, the defect may well be in our own grasp of the argument. Yet in making this very point, Thomason is already defending one of Feyerabend's favourite claims — the reasoning processes used by great scientists are remarkably untidy and do not readily compress themselves into neat philosophic formulation.

When philosophers ask how to assess scientific reasoning, they have generally set themselves high standards. They have sought the best possible answers, justifications defensible against the widest possible range of doubts. For what is the point of devising defences that are evidently weak? As a result, reasoning from 'ELK-facts' (like that employed by Galileo) tends to get overlooked, for it certainly appears to be weak reasoning, and who wants to defend that? Yet such 'ELK-reasoning' is not at all defective *qua* reasoning, and is limited only by its being, its very essence, the drawing of conclusions from a severely limited understanding of the situation to which it is being applied. Yet when confronted with a concrete problem, it may be eminently rational to employ such limited reasoning, for it is poor in one sense only: it does not take us far. It is not poor in the sense of evidently taking us in the wrong direction. If we do not have far to go, reasoning which takes us a short distance may be all we require.

To make this abstract generalisation comprehensible, let us look briefly at a different example of reasoning from ELK-beliefs, a semi-fictional one useful for two reasons: it indicates how commonly we rely of this type of argument; and it removes any suggestion that ELK-facts are particularly tied to the context that Thomason uses to introduce them — assessments of the reliability of scientific instruments.

The computer experts where I work recently flashed a news message over the network advising that a particularly virulent virus had crept onto the network, destroying some files. After a couple of ill-drafted lines of computerese telling us in technical terms what sort of damage had already been detected, we were advised to check our files. Needless to say, this message caused some panic, panic magnified by a number of facts: the computerese was unintelligible to virtually all users; the experts did not clearly tell us how to do what they asked of us; nor did their message tell us which files to check; some users did indeed find serious

P.J. Riggs (ed.), Natural Kinds, Laws of Nature and Scientific Methodology, 235–239.

file damage; others knew they had not made backups; many knew they had very little idea at all about how their computer functioned, etc. etc.

Yet much of the panic was unjustified, in that ELK-understandings (possessed by all but the most stubborn luddites) can reassure us about the immunity of many of our files. We all know, for example, that files are routinely stored via physico-chemical modifications of diskettes; we know that computers do not work by action-at-a-large-distance (in that diskettes need to be inserted into a special device to deliberately change their contents); and we know that they also do not work backwards in time. From all this, we are surely justified in concluding that however ingenious the designer of the virus might be, files stored on diskettes that have not been near a computer since the virus arrived are safe - and in consequence we can formulate an intelligent and pertinent question to pose to the experts: 'How reliably do you know when the virus entered our system?'. If they give a good answer to this question, we can rationally conclude — with great confidence, and despite our extreme ignorance of computer ailments — that a whole category of files are safe from viral attack.

This example also points to another important feature of ELK-facts, something which makes them worth deliberately identifying as a special category: they are the things that specialists usually forget to tell you when you seek their aid. The problem I have in mind here is a very acute one, of considerable economic import: the familiar (but surely strange) fact that experts have so much difficulty in giving instructions to the laity. When I reflect on such examples of this failure of communication that come my way, I find Thomason's analysis of Galileo very helpful in revealing the difficulty. For many of the problems seem caused by the experts forgetting to tell their clients facts analogous to the fact that telescopes work by processing light (or that dinosaurs have ends) - facts critical to understanding what is going on, yet too obvious to the experts to be worth mentioning. I give three example from yet another field: 'In statistics, we use samples because they are small enough to be closely inspected'; 'Statisticians seek to draw inferences about an inaccessible population by looking at relatively accessible samples'; 'The standard deviation is not the only measure of data variability'. It is obviously easy to extend this list.

Though the potency of such ELK-facts is clear, Thomason's discussion tends to blur this fact, and (unintentionally) gives the impression that ELK-reasoning is intrinsically weak. For it is obvious that Galileo's arguments for the reality of Jupiter's moons remain fragile: one cannot possibly draw valid conclusions about the nature of Jupiter from reasoning, no matter how excellent, about the properties of terrestrial telescopes. So Galileo and his contemporaries were genuinely puzzled about the structure of Saturn, and did not insist that the changing views of it through a telescope were as reliable as the evidence for its neighbour's moons.[1] The fact that protuberances were sometimes seen on Saturn was not interpreted as justifying the conclusion that Saturn sometimes had protuberances. Similarly, the fact that Jupiter had something that looked like moons, does not in itself justify the conclusion that Jupiter has moons. That this deduction was in fact made, and plausibly so, reflects not the quality of the direct evidence about the nature of Jupiter, but the status of competing explanations. The appearance of protuberances might well have been caused by something other than temporary ears on Saturn, and observers of Saturn were consciously seeking a plausible hypothesis that could make sense of the 'deceitful' data.[2] But moons were sufficiently familiar to be the preferred mode of explaining the Jovian appearances — once the other alternative, the claim that the appearances were an illusion, and did not even represent a property of Jupiter, was abandoned. In dealing with this alternative hypothesis, the ELK-facts were not at all weak.

This analysis seems to indicate a general *modus operandi* for ELK-reasoning: some hypotheses are so easy to refute, that they fall even to ELK-facts. Yet when scientists are assessing theories they are usually making comparisons between competing theories and (as is widely acknowledged) the evidence which makes them favour a particular theory is so often an assembly of evidence against it competitors. (For it is far easier to attack an hypothesis than to support one, and many hypotheses are in fact readily refuted.) So if a theory is in competition with a hypothesis incompatible with ELK-facts, it can readily by supported by an ELK-fact. Furthermore, if there happens to be nothing in competition with it beyond such an hypothesis, it will be well supported by the ELK-fact.

In Galileo's case, the proponents of the illusion hypothesis were not holding their position because they judged that it made best sense of the actual observations. Rather they seized it hastily, either via too prompt a generalisation from the accepted cases of illusion, or because it offered a refuge from the Galilean alternative and built upon a prejudice that they hoped would find popular appeal. As a result their hypothesis collapsed as soon as it was treated seriously as a scientific hypothesis, for it was incompatible with the most humble of optical facts. As Thomason observes, it was Galileo rather than its proponents who really took the illusion theory seriously. In Galileo's particular case, it was not primarily the reality of the moons of Jupiter that needed defending. What was more pressing here was a relative judgment, about the source of some sense-impressions witnessed via the telescope. Is it better to believe that these are caused by Jupiter? Or better to believe that they are caused by something unrelated to the planet, i.e., are they a complete illusion? Galileo's ELK-arguments are good at showing the illusion hypothesis to be very implausible.[3] Granted then the telescope is telling us *something* about Jupiter, we then have to survey the hypotheses about Jupiter that can account for what is seen in the telescope and in the early 17th Century there was no competitor to the moon theory here. Once the illusion hypothesis had been eliminated, the reality of the moons was accepted by default.

In revealing here the potency of a mode of argument that other philosophers have tended to *denigrate*, Thomason is retracing footsteps, for he did much the same thing in an earlier detection of the role played by what he called ARCHED hypotheses — 'Advanced Risky Conjoined Hypotheses Explaining Data'.[4] When, for instance, Galileo looked at Venus through his telescope, at least three doubtful claims were involved in interpreting the significance of what he saw: (i) the telescope is reliable; (ii) Venus is non-luminous and opaque; (iii) Venus orbits the sun. It might seem that one cannot shed any light on (iii) here, until after one has verified each of (i) and (ii), and to do otherwise is to beg the question etc. Yet Thomason notices that the one set of observations of Venus can give simultaneous evidence for all three doubtful claims: for what is actually seen can be well explained by presuming all three claims to be true. The hypotheses, then, are 'arched' — in the sense of all leaning on each other and giving mutual support.

What these two discussions have in common is very important. They point out firstly, a weakness in much sociology of knowledge (to be elaborated below); and secondly, the dangers of armchair philosophising, i.e., the necessity for students of human thought to 'dirty their hands' with the empirical history of reasoning. For this reveals much that is surprising, complex and unexpected and the student is well advised to attend closely, very closely in fact, to the reasoning processes that clever humans have in fact employed. Since experts (as we have noted) forget about ELK-facts so much of the time, it is particularly difficult to reconstruct reasoning reliably. So hasty or superficial studies of thought run a real risk of being wrong. Neither of Thomason's two examples are in principle beyond the scope of *a priori* investigation, yet both had been declared invalid by acknowledged experts - and both have been resuscitated by closer attention to the historical circumstances. I do not mean to say here that there is anything prescriptive about history: Galileo's arguments are not to be judged valid because of the fact that they once succeeded in convincing people. An assessment of them is still required and many of the arguments used by scientists will surely not survive rigorous scrutiny. But since we are so much better at assessing arguments, than in devising them, history can give us enormous heuristic assistance. It can supply us with modes of argument that we would not have thought of ourselves. For such reasons, philosophers often discuss their work with colleagues. Why not do so with Galileo as well?

More importantly perhaps, Thomason's discussions point to the need to rethink analyses in the Feyerabendian spirit at the foundations of the sociology of knowledge. For Kuhn and Feyerabend are widely believed to have shown that Western Science is far less rational than was once supposed and from this it is concluded that the decisions of science are social in character rather than rational.

I have no difficulty at all with part of this claim, the notion that social causation is vital to the acceptance of scientific theories, and that reason is not universally triumphant. I have indeed elsewhere argued for such an interpretation of the Copernican Revolution.[5] Nor do I find a radical sociology of knowledge to be incompatible with much traditional philosophy of science: I see the standard notion that theories are underdetermined by the evidence, or old-fashioned scrutiny of the complicated conventions implicit in so many scientific theories, as both pointing in this direction.

But I do find myself repeatedly seduced by the charm of high-quality thinking, and I retain my confidence that such thinking is central in the history of science — central in the sense of characterising science, and distinguishing it from other belief-systems. I do not mind admitting the dangers of self-illusion here and would welcome having my errors pointed out to be me if it can be convincingly done in prose of clarity and elegance. To those who think that Feyerabend has shown the way here Thomason has a powerful argument: *Feyerabend was wrong!*

University of Melbourne, Victoria

NOTES

1 Cf. Van Helden (1974), *passim.*
2 Christopher Wren, quoted Van Helden (1974), p.115.
3 This is particularly clear in the case of Saturn. For Galileo agreed that some of the observations were illusions, but insisted that even these illusions were revealing a genuine property of Saturn. See his 'Letter on sunspots', p.102.
4 Thomason (1994).
5 Hutchison (1987) and (1993).

REFERENCES

Galilei, Galileo and Welser, Mark, 'Letters on sunspots' [1613], excerpted & trans. S. Drake, on pp.87-144 of *Discoveries and opinions of Galileo*, ed. S. Drake, (New York: Anchor, 1957 & reprints).

Hutchison, K., 'Towards a Political Iconology of the Copernican Revolution'. On pp.95-141 of *Astrology, Science and Society: Historical Essays*, ed. Patrick Curry, (The Boydell Press: Woodbridge, Suffolk, 1987).

Hutchison, K., 'Harmony and Authority: the Political Symbolism of Copernicus' Personal Seal'. On pp. 115-68 of *Non-verbal communication in science prior to 1900*, ed. Renato Mazzolini, (Florence: Olschki, 1993).

Thomason, N., 'The Power of ARCHED Hypotheses: Feyerabend's Galileo as a Closet Rationalist', *British Journal for the Philosophy of Science* 45 (1994): 255-64.

Van Helden, A., 'Saturn and his Anses', *Journal of the History of Astronomy* 5 (1974): 105-21.

INDEX

AUSTRALIAN STUDIES IN HISTORY AND PHILOSOPHY OF SCIENCE

1. R. McLaughlin (ed.): *What? Where? When? Why?* Essays on Induction, Space and Time, Explanation. Inspired by the Work of Wesley C. Salmon. 1982 ISBN 90-277-1337-5
2. D. Oldroyd and I. Langham (eds.): *The Wider Domain of Evolutionary Thought.* 1983 ISBN 90-277-1477-0
3. R.W. Home (ed.): *Science under Scrutinity.* The Place of History and Philosophy of Science. 1983 ISBN 90-277-1602-1
4. J.A. Schuster and R.R. Yeo (eds.): *The Politics and Rhetoric of Scientific Method.* Historical Studies. 1986 ISBN 90-277-2152-1
5. J. Forge (ed.): *Measurement, Realism and Objectivity.* Essays on Measurement in the Social and Physical Science. 1987 ISBN 90-277-2542-X
6. R. Nola (ed.): *Relativism and Realism in Science.* 1988 ISBN 90-277-2647-7
7. P. Slezak and W.R. Albury (eds.): *Computers, Brains and Minds.* Essays in Cognitive Science. 1989 ISBN 90-277-2759-7
8. H.E. Le Grand (ed.): *Experimental Inquiries.* Historical, Philosophical and Social Studies of Experimentation in Science. 1990 ISBN 0-7923-0790-9
9. R.W. Home and S.G. Kohlstedt (eds.): *International Science and National Scientific Identity.* Australia between Britain and America. 1991 ISBN 0-7923-0938-3
10. S. Gaukroger (ed.): *The Uses of Antiquity.* The Scientific Revolution and the Classical Tradition. 1991 ISBN 0-7923-1130-2
11. P. Griffiths (ed.): *Trees of Life.* Essays in Philosophy of Biology. 1992 ISBN 0-7923-1709-2
12. P.J. Riggs (ed.): *Natural Kinds, Laws of Nature and Scientific Methodology.* 1996 ISBN 0-7923-4225-9

KLUWER ACADEMIC PUBLISHERS – DORDRECHT / BOSTON / LONDON

9 780792 342250